TENASCIN AND COUNTERADHESIVE MOLECULES OF THE EXTRACELLULAR MATRIX

Cell Adhesion and Communication
A series of books encompassing monographs on classes of adhesion molecules and monographs giving a broader functional synopsis on adhesion molecules on a particular system.

Edited by Christo Goridis

Volume 1
Cell Adhesion Molecules in Cancer and Inflammation
edited by A.A. Epenetos and M. Pignatelli

Volume 2
The Laminins
edited by P. Ekblom and R. Timpl

Volume 3
Tenascin and Counteradhesive Molecules of the Extracellular Matrix
edited by K.L. Crossin

Other volumes in preparation

Ig Superfamily of Adhesion Molecules
edited by P. Sondregger

Adhesion Molecules in Lymphocyte Trafficking
edited by A. Hamann

Epithelial Morphogenesis in Development and Disease
edited by W. Birchmeier and C. Birchmeier

This book is part of a series. The publisher will accept continuation orders which may be cancelled at any time and which provide for automatic billing and shipping of each title in the series upon publication. Please write for details.

TENASCIN AND COUNTERADHESIVE MOLECULES OF THE EXTRACELLULAR MATRIX

Edited by

Kathryn L. Crossin

Department of Neurobiology
The Scripps Research Institute
USA

Harwood Academic Publishers
Australia • Canada • China • France • Germany • India • Japan • Luxembourg • Malaysia
The Netherlands • Russia • Singapore • Switzerland • Thailand • United Kingdom

Emmaplein 5
1075 AW Amsterdam
The Netherlands

British Library Cataloguing in Publication Data

Tenascin and counteradhesive molecules of the extracellular matrix — (Cell Adhesion and Communication Series, ISSN 1023-7046; Volume 3)

1. Extracellular matrix 2. Connective tissues
I. Series II. Crossin, Kathryn L.
572.1'9245

ISBN 3-7186-5841-0

CONTENTS

PREFACE TO THE SERIES

The development and normal functioning of all multicellular organisms is governed to a large part by the interactions cells undergo with neighbouring cells and with their acellular environment. Many of these interactions are mediated by cell–cell adhesion molecules and by extracellular matrix components and their cellular receptors, that is by molecules mediating direct cell–cell and cell–matrix contacts. These molecules are particularly important for determining whether a cell remains where it is or moves elsewhere and, if a cell moves, where it goes and when it stops migrating. These are of course key events during normal development, but they play equally crucial roles in adult physiology and pathology, such as the extravasation of white blood cells, inflammatory processes and wound healing, tumour invasiveness and metastasis. Moreover, recent advances in the field show that most, if not all cell adhesion molecules are capable of triggering intracellular events, in the same way as diffusible growth and differentiation factors and their cellular receptors do. It is thus hardly surprising that clinicians are devoting increasing attention to the molecular mechanisms underlying cell adhesion, and that cell adhesion molecules are now being considered as suitable targets for drug development.

This book series is aimed at scientists, both in academia and in industry, and at graduate students planning to move into the area, at the clinician, who wants to become familiar with a field with many clinical implications, at scientists already working in the field, who want to keep abreast with the recent developments outside their own speciality. Hence, each volume of the series provides a review of a particular segment of the field as well as critical assessment of recent discoveries and future developments. Each volume has a volume editor, who invites contributions to cover the different aspects of the topic. By keeping the number of contributors small, we hope to avoid overlaps and redundancies, common pitfalls of multi-author volumes.

One part of the series will concentrate on important structural families of molecules, another one will address the role of cell adhesion in selected physiology and pathological phenomena. The previous volumes *The Laminins* and *Cell Adhesion Molecules in Cancer and Inflammation* are excellent examples of the first and second kind, respectively. The present volume deals both with an important class of extracellular matrix components, the tenascins, and with a biological phenomenon whose widespread occurrence has been recognized only recently, and which has been termed counter-adhesion or repulsion. In fact, transient contacts with extracellular matrix components may result in two types of cellular response, the establishment of more or less stable contacts or an avoidance reaction. Often, the same kind of extracellular matrix molecule may be involved in both phenomena. Tenascin was the first molecule shown to mediate counteradhesive effects, hence the space devoted to it in this volume, but other extracellular matrix molecules are also dealt with.

I am indebted to the staff at Harwood Academic Publishers, who made this enterprise possible, to the volume editors, each one an authority in his/her field, and last but not least to the many competent contributors.

Christo Goridis

INTRODUCTION

A fundamental question in developmental biology is to understand the complex relationship between gene expression and morphogenesis. A large body of evidence over the last decade has implicated molecules mediating cell adhesion as critical players guiding the cellular events that lead to the emergence of form in developing embryos. These molecules are also essential to the maintenance of form in adult animals, and their functions may be perturbed in disease and degenerative states. Because of their role in guiding these morphological processes, cell–cell adhesion molecules, cell–substrate adhesion molecules, and cell–junctional molecules have been termed morphoregulatory molecules. Harwood Academic Publishers' book series on Cell Adhesion recognizes the importance of the different types of adhesion molecules and provides a unique opportunity to examine the various aspects of the cell and molecular biology of adhesion molecules.

This book is concerned with the structure, function, and regulation of a fascinating extracellular matrix (ECM) molecule called tenascin. It is a large multidomain glycoprotein with dynamic expression patterns and dramatic effects on cell morphology and physiology. Various molecules of the ECM have long been known to affect cellular processes, which in turn can influence morphogenesis. ECM components interact with each other and with cell surface receptors resulting in the adhesion and polarity of epithelial tissues During development, they are found in areas of extensive cell migration. They are known to affect cell proliferation and differentiation and to result in changes in gene expression in cells encountering different extracellular matrix components. Together with cell–cell adhesion molecules, they are important modulators of neurite outgrowth and guidance.

Tenascin was independently discovered in a number of different laboratories utilizing a variety of different assays. For example, it was discovered as a molecule present in mesenchymal cells and in glioma cells at much higher levels than in normal adult tissue, and was therefore designated the glioma-mesenchymal extracellular matrix protein (GMEM). It was also discovered using antibodies that showed it to be a molecule enriched at the myotendinous junction and was termed the myotendinous antigen. In addition, it played a role in mediating neuron-glia adhesion, inasmuch as antibodies that recognized the protein were able to inhibit the binding of neurons to glial monolayers. In this discovery, the molecule was called cytotactin, in recognition of its role in mediating cellular interaction. The same molecule was also described as an extracellular matrix protein involved in tissue interactions during both development and oncogenesis and was termed tenascin. Tenascin was found as a contaminant of fibronectin preparations that showed a distinctive appearance in rotary shadowing electron micrographs and, because of its now characteristic structure, was termed hexabrachion. In other studies, the protein was identified as one of the soluble brain molecules that carried the L2/HNK1 carbohydrate antigen and was labeled J1. Tenascin is a member of a growing family of molecules that have similar regions of homology to epidermal growth factor repeats, fibronectin type III repeats, and fibrinogen. These include restrictin/janusin and tenascin-X/tenascin-MHC. A recently adopted nomenclature refers to these three TN family members as TN-C, TN-R, and TN-X. Whether these proteins share biological functions with tenascin is a subject of intense interest that is discussed in this volume.

In recent years, it has become clear that counteradhesive forces (those that counterbalance cell and substrate adhesion) are also important in structuring the embryo particularly in the development of the nervous system. Counteradhesive forces have been shown to be mediated by a number of molecules—tenascin, the major subject of this monograph—as well as several other extracellular matrix molecules including thrombospondin, SPARC and chondroitin sulfate proteoglycans. Counteradhesive activity appears to be particularly important in regulating cell morphology, inhibiting cells from adhering too tightly to other extracellular matrix proteins and from flattening on these substrates. Through interactions with cell surface receptors, notably integrins, counteradhesive molecules may affect the regulation of intracellular pH and may subsequently affect cell proliferation and migration and neurite sprouting and elongation in the nervous system.

The phenomenon of counteradhesion has been postulated to be an important process mediated by tenascin and the other molecules discussed in this monograph that are found in areas of extensive tissue remodelling during development, wound healing, and regeneration. Further understanding of the molecular mechanisms underlying counteradhesive activity will shed light on these fundamental processes. The knowledge gained should help in determining how these processes might be manipulated to alter various aspects of embryonic development, to enhance regeneration, and to affect the outcome of a variety of diseases.

CONTRIBUTORS

James A. Bassuk
Department of Biological Structure
SM20
University of Washington
Seattle WA 98195
USA

Laura Borsi
Laboratory of Cell Biology
Istituto Nazionale per la Ricerca
sul Cancro
Viale Benedetto XV, 10
16132 Genoa
Italy

Barbara Carnemolla
Laboratory of Cell Biology
Istituto Nazionale per la Ricerca
sul Cancro
Viale Benedetto XV, 10
16132 Genoa
Italy

Donald W. Copertino
Department of Neurobiology
The Scripps Research Institute
10666 North Torrey Pines Road
La Jolla CA 92037
USA

Kathryn L. Crossin
Department of Neurobiology
The Scripps Research Institute
10666 North Torrey Pines Road
La Jolla CA 92037
USA

Andreas Faissner
Department of Neurobiology
University of Heidelberg
INF 364
6900 Heidelberg
Germany

Frederick S. Jones III
Department of Neurobiology
The Scripps Research Institute
10666 North Torrey Pines Road
La Jolla CA 92037
USA

Jack Lawler
Harvard Medical School
Department of Pathology,
Brigham Women's Hospital
LMRC414 221 Longwood Avenue
Boston MA 02115
USA

Renée K. Margolis
Department of Pharmacology
State University of New York
Health Science Center
Brooklyn NY 11203
USA

Richard U. Margolis
New York University Medical Center
Department of Pharmacology
550 First Avenue
New York NY 10016
USA

Vincent P. Mauro
Department of Neurobiology
The Scripps Research Institute
10666 North Torrey Pines Road
La Jolla CA 92037
USA

Kouros Motamed
Department of Biological Structure
SM20
University of Washington
Seattle WA 98195
USA

Anne L. Prieto
Department of Neurobiology
The Scripps Research Institute
10666 North Torrey Pines Road
La Jolla CA 92037
USA

E. Helene Sage
Department of Biological Structure
SM20
University of Washington
Seattle WA 98195
USA

Annalisa Siri
Laboratory of Cell Biology
Istituto Nazionale per la Ricerca sul Cancro
Viale Benedetto XV, 10
16132 Genoa
Italy

Luciano Zardi
Laboratory of Cell Biology
Istituto Nazionale per la Ricerca sul Cancro
Viale Benedetto XV, 10
16132 Genoa
Italy

1. THE MOLECULAR BIOLOGY OF TENASCIN: STRUCTURE, SPLICE VARIANTS, AND REGULATION OF GENE EXPRESSION

FREDERICK S. JONES* and DONALD W. COPERTINO

Department of Neurobiology, The Scripps Research Institute, 10666 North Torrey Pines Road, La Jolla, California 92037

INTRODUCTION

Tenascin (TN) is an extracellular matrix (ECM) glycoprotein with a highly symmetrical hexameric structure called a hexabrachion. Tenascin polypeptides are modular in structure and a number of variants can be produced through alternative splicing of primary transcripts. Tenascin, like other proteins of the ECM, binds to many different ligands which include the integrin family of cell surface receptors, ECM proteins such proteoglycans, fibronectin, and collagen, and small molecules such as glycolipids. Tenascin appears in a site-restricted fashion during embryonic development. It is also re-expressed in the adult during tumorigenesis and wound healing. In this chapter, we will review the structure of the TN molecule, the variants produced by alternative splicing of RNA, and discuss recent studies which have identified cis-regulatory elements and the trans-acting protein factors that regulate TN gene expression.

A DETAILED STRUCTURAL MODEL OF TENASCIN

Tenascin was discovered independently by a number of laboratories studying different aspects of cell biology. Accordingly, it has a number of names including GMEM (Bourdon *et al.*, 1983), myotendinous antigen (Chiquet and Fambrough, 1984), cytotactin (Grumet *et al.*, 1985), J1220/200 (Kruse *et al.*, 1985), hexabrachion (Erickson and Iglesias, 1984), and neuronectin (Rettig *et al.*, 1989). Tenascin as isolated from the extracellular matrix of many different tissues is a six-armed structure called a hexabrachion. This structure was first visualized in electron micrographs of rotary shadowed preparations of ECM isolated from human fibroblasts (Erickson and Iglesias, 1984). Tenascin hexabrachions (Figure 1.1, *Top*) have remarkable symmetry. Six arms radiate from a central core. The proximal portions of the arms are thin but at a discrete point thicken and then terminate in a globular particle. Early

* Corresponding author

biochemical studies (Chiquet and Fambrough, 1984; Grumet *et al.*, 1985; Kruse *et al.*, 1984; Grumet *et al.*, 1985; Friedlander *et al.*, 1988; Hoffman *et al.*, 1988) indicated TN was a disulfide-linked oligomer with subunits ranging from 190–300 kilodaltons (kDa). Based on proteolytic mapping studies and cDNA cloning experiments for chicken TN, a detailed model of the hexabrachion has been proposed (Jones *et al.*, 1988; Pearson *et al.*, 1988; Jones *et al.*, 1989; Spring *et al.*, 1989) (Figure 1.1, *Bottom*). This model is based on similarities to other proteins, the orientation and interaction of the polypeptides, and the elements that correspond to features seen in electron micrographs.

Tenascin polypeptides are divided into four segments: an amino-terminal region containing many cysteine residues which participate in oligomer formation, a series of compact epidermal growth factor-like (EGFL) repeats, an array of fibronectin type III repeats, some of which can be included or excluded by differential RNA splicing events, and a carboxyl-terminal segment that resembles the distal domain from the beta and gamma chains of fibrinogen. We would like to briefly elaborate on these four separate regions of TN (using the chicken molecule as the exemplar) to understand better the overall structure of the hexabrachion and its relationship to proposed functions for the molecule (see Chapter 2). Two other proteins contain all of these domains, TN-R (also known as Janusin, $J1_{160/180}$, and restrictin (Fuss *et al.*, 1993; Norenberg *et al.*, 1992), and TN-X (also known as TN-MHC, XB, and p450opp (Morel *et al.*, 1989; Bristow *et al.*, 1993) suggesting that a number of different TN family members may exist.

The first 23 amino acids of TN as predicted by the cDNA sequence encode a typical signal sequence followed by a 10 amino acid basic region similar to precursor sequences of other adhesion molecules. The amino terminus of the mature TN polypeptide as isolated from chicken brain was determined directly by protein sequencing. It mapped to a glutamate residue located 34 amino acids downstream from the initiator methionine (Jones *et al.*, 1989). The amino terminal region of TN contains eight cysteine residues and 3–4 heptad repeats (Pearson *et al.*, 1988; Jones *et al.*, 1989; Spring *et al.*, 1989). The cysteine residues are thought to link individual polypeptides into oligomeric structures. The heptad repeats form an alpha helix which allow three TN monomers to interact in a parallel triple-stranded coiled-coil and form trimers. TN trimers are thought to be further stabilized covalently by disulfide bonds. At least two cysteine residues from each TN polypeptide chain (6 total) would be required to link three chains together in a disulfide ring (Erickson and Bourdon, 1989). Two such rings form on either side of the coiled-coil region of fibrinogen (Doolittle, 1984). Such interactions require the involvement of four cysteine residues from each chain (2 on either side of the alpha helical region) in 12 disulfide linkages. Thus, it is reasonable to propose that 12 disulfide linkages stabilize each TN trimer (6 in each disulfide ring flanking the heptad region). Two different alignments have been proposed for the amino acids within the heptad repeats. One positions three cysteine residues

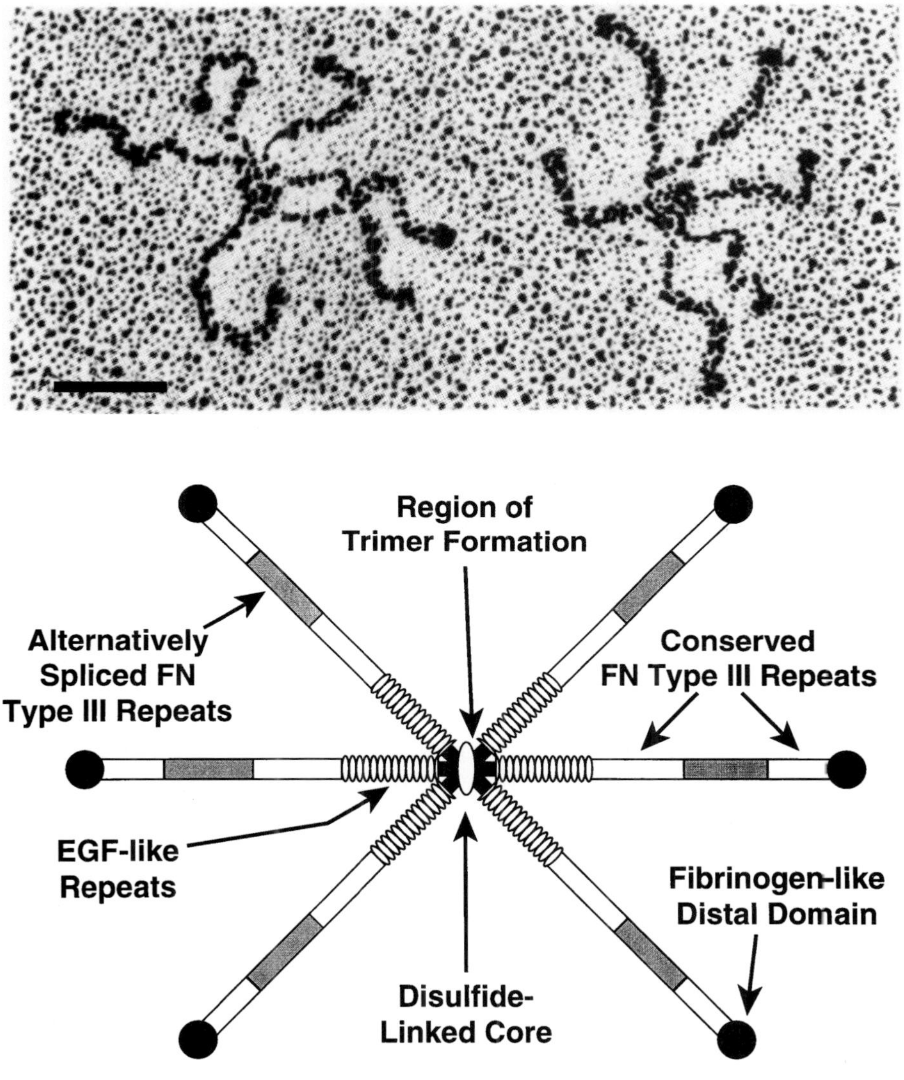

Figure 1.1 A model for the tenascin hexabrachion. *(Top)* Electron micrograph of rotary shadowed tenascin isolated from the ECM of mouse fibroblasts (Jones, 1991). Bar, 50 nm. *(Bottom)* Schematic representation of the assembly of six tenascin polypeptides into the hexabrachion. The oval at the center represents the site for the linkage of two trimers through interchain disulfide bonds. The black regions extending from the central core correspond to the region of trimer formation. This region of the tenascin polypeptide contains many cysteine residues and a series of heptad repeats. The smaller ovals refer to the rigid epidermal growth factor (EGF)-like repeats; the unfilled and shaded rectangles refer to the constant and variable fibronectin (FN) type III repeats, respectively; and the black globular domains at the ends of each polypeptide represent the region with homology to the distal domains of the beta and gamma chains of fibrinogen.

within the last heptad repeat (Spring *et al.*, 1989), whereas the other places these cysteine residues outside of the heptad array (Conway and Parry, 1991). It has also been proposed that disulfide bonds link TN trimers into hexamers (Friedlander *et al.*, 1988; Jones *et al.*, 1989; Spring *et al.*, 1989). Tenascin polypeptides have also been shown to form larger aggregates. For example, a TN-like molecule from the porifera *Oscarella tuberculata* forms aggregates containing 2–6 hexamers (Humbert-David and Garrone, 1993). Thus, it is unknown how many disulfide linkages are involved in the formation of trimers, hexamers, and higher order aggregates. It is also possible that other small linker molecules play a role in the oligomerization of TN polypeptides.

Proceeding outward from the central core, the next domain is formed by epidermal growth factor-like (EGFL) repeats. Each repeat is 31 amino acids in length and contains six cysteine residues which are capable of intrachain disulfide linkages. The TN EGFL repeats are exceptionally compact in nature. Unlike the EGF-like repeats found in other proteins such as coagulation factors IX and X (Handford *et al.*, 1990), the TN repeats lack the acidic residues required for binding of calcium. The TN repeats also do not contain a large beta strand in the center of the repeat, as is found, for example, in epidermal growth factor (Cooke *et al.*, 1987). The compact nature of the TN EGF-like repeats may provide rigidity to the polypeptide chain – a specific structural feature that may be responsible for the highly symmetrical radiation of TN arms from the central core of the hexabrachion. As will be discussed in more detail in Chapters 2 and 3, the EGF-like repeats have been found to affect a rounded morphology and induce neurite retraction in cells that interact with this region of TN (Spring *et al.*, 1989; Prieto *et al.*, 1992; Crossin, 1994).

A large portion of the TN polypeptide is made up of a linear array of fibronectin type III repeats. These are composed of 91 amino acids and map onto the distal portions of the hexabrachion arms. Type III domains have an extended beta structure allowing the distal region of TN to be highly flexible and particularly susceptible to protease digestion. Recently, the crystal structure has been determined for the third fibronectin type III domain of TN (Leahy *et al.*, 1992). It is a globular structure assembled as a sandwich of seven anti-parallel beta strands arranged in two sheets (Leahy *et al.*, 1992; Patthy, 1990). The type III domain has many features in common with a C-type immunoglobulin domain. However, unlike the Ig domains, the type III domain lacks the two cysteine residues which provide the intrachain disulfide bonds that link pairs of beta sheets together. The overall folding topology of the type III domain from TN is essentially identical to that from a similar domain from fibronectin (Baron *et al.*, 1992). While type III domains appear to share a common secondary structure, their amino acid composition can show great variability. For example, the type III domains of TN are between 26–40% identical to those from fibronectin (Jones *et al.*, 1988). Specific peptide sequences (such as RGD) endow individual type III domains with discrete binding functions. Thus, the assembly of unique type III domains into a repetitive array creates diversity and multivalence. This allows a particular array which is characteristic of a specific extracellular matrix or cell surface protein to interact with a unique set of ligands. The composition of type III array in TN can be altered by alternative splicing of RNA in a segment which corresponds to the region between the fifth and sixth fibronectin type III repeats (Jones *et al.*, 1988; Jones *et al.*, 1989; Spring *et al.*, 1989). Alternative splicing generates further structural and perhaps functional diversity for TN by producing a number of different isoforms of the protein. A detailed analysis of the different splice variants of TN found in a number of vertebrate species is discussed below (see Figure 1.2 and Table 1.1).

The most carboxyl-terminal domain of TN forms the terminal knob of the hexabrachion. This region of TN is approximately 32% identical to a corresponding distal domain segment found in the beta and gamma chains of fibrinogen (Doolittle, 1984). As observed in fibrinogen, the distal domain has two prominent polypeptide loops formed by two consecutive intrachain disulfide linkages from four cysteine residues. The distal domain as found in the TN sequence preserves the exact dimensions in spacing of the two interchain disulfide loops. The polypeptide segment between these disulfide loops forms an EF-hand calcium binding site (Tuffy and Kretsinger, 1975).

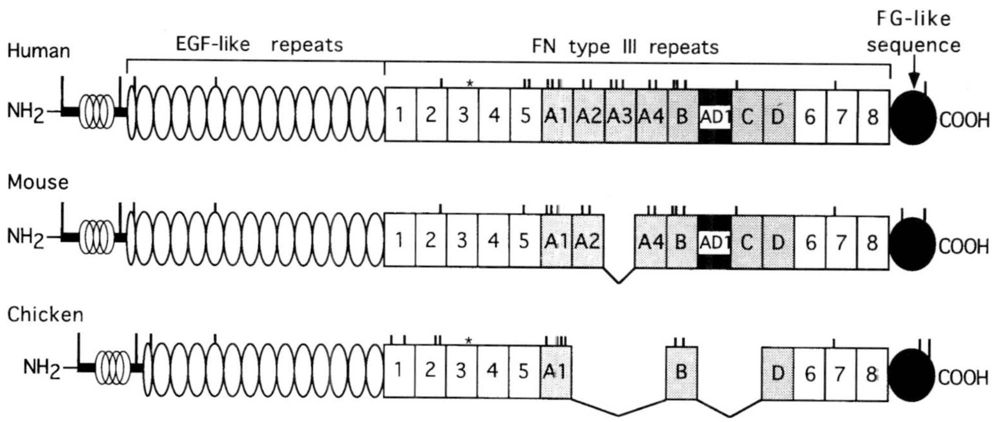

Figure 1.2 Schematic representation of the domain structure of the human, mouse and chicken tenascin polypeptides. The EGF-like repeats, the fibronectin (FN) type III repeats, and the globular domain with similarity to the beta and gamma chains of fibrinogen (FG) are indicated. The cysteine-rich regions at the amino terminal segment of the polypeptides are indicated by thick horizontal lines. The heptad repeats within the amino terminal region are depicted by small overlapping ovals. The differentially spliced type III repeats (shaded) of mouse and chicken are aligned according to sequence similarity to the human repeats. The asterisks mark the RGD tripeptide in the third FN type III repeat in the human and chicken molecules. Potential N-glycosylation sites are indicated with thin vertical lines.

Table 1.1 Tenascin mRNAs generated by alternative splicing

Species	*Number of variants*	*FN type III repeats present*
Chicken	3	1–5, A1, B, D, 6–8 1–5, D, 6–8 1–8
Pig	3	1–5, B, D, 6–8 1–5, D, 6–8 1–8
Mouse	7	1–5, Al, A2, A4, B, C, D, 6–8 1–5, A1, A2, A4, B, D, 6–8 1–5, A1, A2, A4, D, 6–8 1–5, A1, A2, 6–8 1–5, A1, 6–8 1–5, D, 6–8 1–8
Human	8	1–5, Al, A2, A3, A4, B, AD1, C, D, 6–8 1–5, A1, A2, A3, A4, B, C, D, 6–8 1–5, A1, A2, A3, A4, B, AD1, 6–8 1–5, A1, A2, A3, A4, B, D, 6–8 1–5, A1, A2, A3, A4, B, 6–8 1–5, B, D, 6–8 1–5, D, 6–8 1–8

The EF-hand consists of an alpha helix, a loop around the calcium ion, and another alpha helix. Tenascin contains an EF-hand calcium binding site with four out of the six vertices of the tetrahedral calcium coordination complex occupied by an oxygen-containing residue. Indeed, we have shown in unpublished data that native TN isolated from chick brains binds $^{45}Ca^{2+}$. It is likely that binding of calcium in this region of TN may determine a specific conformation which is important for its function.

Recently, a number of fibrinogen-like sequences have been characterized in a number of molecules. Interestingly, all are similar to the distal domain. Additional members of the extended fibrinogen distal domain family include: (1) other members of the TN family TN-R and TN-X ; (2) PT49, a cytotoxic T cell-specific cDNA isolated by subtractive hybridization (Koyama *et al.*, 1987); (3) FRePA, a cDNA clone isolated from a library prepared from the soft tissue of the sea cucumber *Parastichopus parvimensus* (Xu and Doolittle, 1990); and (4) the product of the *scabrous* gene from *Drosophila*, which is important for eye development (Baker *et al.*, 1990). Evolutionary studies (Xu and Doolittle, 1990; Jones, 1991) predict that a primordial distal domain duplicated and diverged into two primary lineages, one giving rise to the fibrinogens and another leading to the TNs and other molecules such as PT49, FRepA, and scabrous. This divergence occurred prior to separation of the beta and gamma chains and well before the emergence of fibrinogen-based blood clotting and the vertebrate lineage, which occurred 350–400 million years ago.

In fibrinogen, the distal domain participates in protein associations required for fibrin polymerization and clot formation. However, little is known about the function of the fibrinogen-like distal domain in TN or in other proteins in which it is found. The distal domain in TN and other molecules may participate in similar protein-protein interactions similar to those observed for fibrinogen. Indeed the terminal knobs of hexabrachions appear to interact with collagen fibrils in electron microscopic analyses of vitreous preparations (Wright and Mayne, 1988) and it has been shown that particular forms of TN interact with collagen (Probstmeier *et al.*, 1990). A short peptide motif within the distal domain of β-fibrinogen has been shown to bind to the ECM protein thrombospondin (Bacon-Baguley *et al.*, 1987). A nearly identical sequence is present in TN.

TENASCIN ISOFORMS GENERATED BY ALTERNATIVE SPLICING

The expression of different isoforms of tenascin can be attributed to alternative splicing in the mRNA. Expression of these variants depends on the stage of development, tissue, cell-type, and cell line (Chiquet-Ehrismann *et al.*, 1991). The characterization of multiple TN polypeptides from different species has been deduced from the isolation of cDNA clones. The first five and the last three fibronectin type III domains of TN are present in all cDNA clones characterized to date. All variants are produced by alternative splicing of primary RNA encoding the type III repeats between the fifth and sixth repeats (Figure 1.2).

The finding that multiple transcripts encoded the different isoforms of TN was first inferred by northern hybridization experiments in the chicken system. Early studies showed that at least 5 different TN mRNAs between 6.4 and 8.3 kb were present in the chick between days 9–15 of embryonic development (Jones *et al.*, 1988; Jones *et al.*, 1989). A northern analysis using a cDNA probe to the constant region of chicken TN detected transcripts of 7.2 and 6.4 kb at embryonic day 15 (E15) in the brain and 8.3, 8.0 and 6.4 kb transcripts in smooth muscle (gizzard). A cDNA probe specific for the differentially spliced repeats B and D hybridized to the 7.2 kb species in brain and the 8.3 and 8.0 kb species in gizzard. The 7.2 kb RNA was subsequently shown to encode the 220 kDa form of TN since only these species were detected in isolated cerebella at hatching and postnatal day 3 (P3)(Prieto *et al.*, 1990). It is likely that the 8.0 and 8.3 kb transcripts contain additional type III repeats (see below).

The entire structure of the chicken TN molecule has been determined by cDNA cloning of three splicing variants (Jones *et al.*, 1988; Pearson *et al.*, 1988; Jones *et al.*, 1989; Spring *et al.*, 1989) (Figure 1.2, Table 1.1). The largest cDNA contains three additional repeats (A1, B, and D) for a total of 11. This cDNA clone encodes 6100 base pairs (bp), 234 bp of 5′ untranslated sequence, 439 bp of 3′ untranslated sequence, and 5424 bp of coding region. Therefore, the large isoform of TN contains 1808 amino acids (Mr=199,000). A cDNA encoding the shortest TN molecule lacks the three type III repeats following the fifth repeat, and contains 1535 amino acids (Mr=169,000). Based on the molecular mass of type III repeats these two cDNAs encode the 220 kDa and 190 kDa forms. Another splicing variant of intermediate size contains 1626 amino acids (Mr=179,000) with only one additional repeat (D) and encodes the 200 kDa variant.

The expression of TN splicing variants changes during development in the chick embryo. Using probes which recognized either all forms of TN or only the variable spliced repeats B and D, it was shown by *in situ* hybridization that the TN splice variants are expressed differentially in tissues and cell types during development (Prieto *et al.*, 1990; Tucker and McKay, 1991). Using monoclonal antibodies to the various forms of chicken TN, Chiquet-Ehrismann *et al.* (Chiquet-Ehrismann *et al.*, 1991) showed the differential segregation of TN isoforms in cell types of the developing chicken gizzard with the gradual loss of the large 220 (230) kDa form from gizzard smooth muscle late during development.

Multiple transcripts, cDNA clones, and protein isoforms have also been described for mouse, human, and pig TN (Weller *et al.*, 1991; Gulcher *et al.*, 1989; Nishi *et al.*, 1991). In the developing mouse gut, the 210 kDa form is found predominately early while the 260/240 kDa form appears late (Aufderheide and Ekblom, 1988). Using a cDNA clone to a constant region of the molecule as a probe in northern hybridization experiments, two transcripts of approximately 8 kb and 6 kb were detected (Weller *et al.*, 1991). These two transcripts were shown to be differentially expressed during development. The 8 kb species was more prominent in the prenatal kidney, whereas the 6 kb transcript predominated in the developing gut. The correspondence between the northern and protein data in the developing gut suggested that the 8 kb transcript encodes the 260/240 kD form and the 6 kb transcript encodes the 210 kDa variant.

The first two splicing variants cloned for mouse TN contained 8 and 13 type III repeats (Weller *et al.*, 1991). The longer cDNA contains 6831 bp, 126 bp of 5′ untranslated sequence, 645 bp of 3′ untranslated sequence, and 6060 bp of coding region. The larger cDNA contains 2019 amino acids (Mr=222,000) and the smaller variant contains 1564 amino acids (Mr=172,000). Treatment of the large form of TN with endoglycosidase F reduced the apparent molecular mass of this molecule from 260 kDa to approximately 230 kDa, which is in close agreement with the size predicted by the cDNA sequence. Ten of the 19 N-glycosylation sites for the larger form of mouse TN are within the alternatively spliced segment (Figure 1.2).

Four TN polypeptides with molecular weights 240, 225, 210, and 190 kDa have been detected by western analysis in the developing mouse brain (Faissner and Kruse, 1990; Bartsch *et al.*, 1992). A developmental analysis of mouse brain mRNAs using a PCR amplification of cDNAs (RT-PCR) strategy has detected four major splice variants with six (A1, A2, A4, B, C, D), five (A1, A2, A4, B, D), one (D) or no alternatively spliced type III repeats whose expression correlate with the TN polypeptides (Dörries and Schachner, 1994)(Table 1.1). In addition, another type III repeat homologous to the human AD1 repeat was identified in mouse genomic DNA (Sriramarao *et al.*, 1993) (Figure 1.2). This repeat has not been detected within mouse TN mRNAs. Therefore, other splice variants and combinations of repeats are possible, some of which are known to exist (Table 1.1) and could account for the different TN isoforms.

Human TN polypeptides of 240 and 180 kDa have been detected by western analysis (Gulcher *et al.*, 1989). A northern analysis using a cDNA probe to a constant region detected transcripts of 7.8 kb and 6.2 kb, and a probe directed against the alternatively spliced region detected only the 7.8 kb transcript (Gulcher *et al.*, 1989). Initially, cDNA cloning of human TN transcripts identified two splicing variants with a total of 8 and 15 type III repeats (Gulcher *et al.*, 1989). However, a more sensitive analysis using RT-PCR has identified several more splicing variants (Siri *et al.*, 1991)(Table 1.1). A cDNA clone was isolated in which only two of the seven alternatively spliced repeats were present (repeats B and D). Another cDNA clone contained six of the seven repeats with only repeat C being removed. This analysis also detected an isoform that contained all seven alternatively spliced repeats. However, the cDNA that lacks these repeats was not detected. A more recent analysis of cDNA clones from a library prepared from human U251MG glioblastoma cells has revealed another repeat between repeats C and D designated AD1 (Sriramarao *et al.*, 1993) (Figure 1.2, Table 1.1). Three splice variants have also been described for porcine TN in which eight (1–8), nine (1–5, D, 6–8) and ten (1–5, B, D, 6–8) type III repeats are present (Nishi *et al.*, 1991)(Table 1). It seems likely that other TN cDNAs also exist, not only for human, but for other species as well. A great number of TN variants may exist in each species if all combinations are possible.

Alternative splicing of pre-mRNAs provides a mechanism for generating multiple functional proteins from a single gene (for reviews, see Smith *et al.*, 1989; McKeown, 1992). Alternative splicing is often regulated in a developmental or tissue-specific manner. Regulation at the level of splice site selection involves the recognition of cis-encoded elements within the pre-mRNA by trans-acting factors. Little is known about how the alternative splicing of TN is regulated. A recent report shows that the pH of the culture medium has a dramatic effect on the splicing pattern of TN mRNAs (Borsi *et al.*, 1995). Likewise, the steady-state levels of human TN splice variants change in response to agents that change the proliferative state of the cell (Borsi *et al.*, 1994). The mRNA encoding the large form is more prominent in proliferative cells than quiescent cells. Clearly the cellular environment greatly affects the generation of TN splice variants either by altering the levels of splicing factors or interactions between splicing factors and the pre-mRNA.

The functional consequence of the different splicing variants of TN is unclear. As described above, it is clear that splice variants of TN are differentially expressed during development. In some developmental systems, the large form of TN is expressed early and is lost later during development, whereas in other systems the large form is expressed late. The large variant of TN seems to be expressed at times when cellular processes require active cell migration or tissue remodelling. The expression of the chicken 220 kD form of TN correlated with embryonic cell migration and pattern formation in the cornea (Kaplony *et al.*, 1991). As the cornea became fully differentiated and there was an increase in synthesis of the ECM by corneal fibroblasts, the expression of the 220 kD form was reduced. The large isoform of TN is also reexpressed during the initial stages of wound healing in the adult skin but disappears following healing (Chuong and Chen, 1991). In addition, the large splice variant of TN mRNA is ten times more abundant in invasive breast carcinomas than in normal breast tissue (Borsi *et al.*, 1992). Recently, annexin II localized at the cell surface has been reported to be a putative receptor for the alternatively spliced repeats for human TN A1–A4, B, C, and D (Chung and Erickson, 1994). It has also been reported that the smallest TN variant in chicken has a higher affinity for fibronectin than the other forms, and that this variant is the only form expressed in tissues rich in extracellular matrix (Chiquet-Ehrismann *et al.*, 1991). Thus differential expression of TN variants may select for the binding of specific cell types or ECM components. Nevertheless, the functional consequences of the selection of a particular TN variant over another have not been demonstrated and the mechanisms that underlie preferential splice site selection are currently unknown.

REGULATION OF TENASCIN EXPRESSION

Tenascin is synthesized by a wide variety of neural and non-neural cells during embryogenesis. The dynamic and transient nature of TN expression during development distinguishes TN from other ECM glycoproteins, such as fibronectin and laminin. Tenascin expression is coordinated in both time and space. This property coupled with the effects of the molecule on cell behavior suggests that TN could orchestrate tissue patterning by determining whether cells adhere to the substratum or to themselves, thereby altering their migration routes and patterns of process extension. Thus, a key goal in studying TN biology is to determine the molecules and the regulatory pathways that govern the local expression of TN.

During development of the chicken embryo, TN is first expressed in anterior regions of the embryo during gastrulation and somite formation and in each case expression advances towards the tail (Crossin *et al.*, 1986). A particularly striking example of these waves of expression is seen in somites whose origin in the chick embryo is a posterior structure of the segmental plate. Pattern of TN synthesis in the somites reflects the distinct morphological stage of somitogenesis which itself appears to depend on the relative position of the mesoderm and the anterior/posterior axis of the embryo. It is thus a reasonable hypothesis that the regulation of TN gene expression could be controlled by transcription factors that have a graded cephalocaudal distribution. At a more advanced stage of somitogenesis, TN is expressed in the rostral half of the somite concurrent with the invasion of neural crest cells (Tan *et al.*, 1987). This localized distribution of TN in the somite represents a refinement of pattern that is reminiscent of the assignment of segment polarity in *Drosophila melanogaster* development (McGinnis and Krumlauf, 1992).

At later stages of development, TN is expressed in the central and peripheral nervous system by glia in complex patterns that correlate with neuronal differentiation and migration events. At non-neural sites, TN is produced at inductive sites and during morphogenesis of a number of organs. For example, TN is expressed in narrow sheets of dense mesenchyme that surround budding epithelia or growing bronchioles of the lung (Crossin *et al.*, 1986; Prieto *et al.*, 1990). It is also actively synthesized by perichondroblasts and in the periostea of the limbs in patterns that prefigure and outline the exact shape of the bone that will develop (Prieto *et al.*, 1990). These patterns of TN synthesis reflect temporal growth gradients that are an important part of morphogenesis and suggest that control of the gene expression is undoubtedly complex, perhaps being governed at many levels depending on the cell type, state, and position in the developing embryo.

INDUCTION OF TENASCIN BY GROWTH FACTORS

Growth factors that are secreted by actively proliferating cells can induce synthesis of tenascin in neighboring cells. Chiquet-Ehrismann *et al.* (Chiquet-Ehrismann *et al.*, 1986) found that adult rat mammary gland fibroblasts were repressed from synthesizing TN *in vivo* while they could produce the molecule in abundance when cultured in the presence of fetal bovine serum. Tenascin was found also to be produced by fibroblasts in the stroma surrounding mammary carcinomas (Inaguma *et al.*, 1988). These observations suggested that the production of TN-rich matrices surrounding neoplastic epithelia may correlate with tumor invasiveness and metastasis. These studies prompted the investigation of whether specific growth factors induce the expression of TN.

Early studies showed that both fetal bovine serum and TGF-β could increase the levels of TN produced by chicken embryo fibroblasts (Pearson *et al.*, 1988). A number of neural tumor cell lines have been distinguished by virtue of their patterns of TN synthesis in the presence of different growth factors. The U251MG glioma cell line constitutively expresses TN in the absence of growth factors (Erickson and Bourdon, 1989). However, a number of cell lines derived from undifferentiated neural tumors alter their TN levels in the presence of growth factors. Acidic FGF, basic FGF, and the phorbol ester TPA stimulate TN secretion from the cell line derived from a pediatric neural tumor, Ewings sarcoma, while only TPA could stimulate secretion from the LAI55 cell line derived from a neuroblastoma (Rettig *et al.*, 1989). This transient expression of TN could be abolished with cycloheximide and actinomycin D treatment, indicating that the induction required *de novo* protein and RNA synthesis (Rettig *et al.*, 1989). Furthermore, these data suggested that induction of TN mRNA synthesis by FGF and TPA may occur through signal pathways that converge onto the components of the transcriptional machinery assembled on the TN promoter.

ISOLATION AND CHARACTERIZATION OF THE REGULATORY SEQUENCES FOR TENASCIN

As a first step toward better understanding the molecular components that direct tenascin gene expression patterns, the promoter region of the chicken TN gene was isolated and characterized. Early studies on the chicken TN promoter revealed that a 4.3 kilobase segment was sufficient to control constitutive expression of a CAT reporter gene in chicken embryo fibroblasts (Jones *et al.*, 1990). The same segment directed cell specific expression of TN in U251MG cells but not in HT1080 cells which do not normally express the molecule. A number of putative regulatory elements were identified in this region of the promoter, including response elements for growth factor signal transduction and sequences resembling homeodomain binding sites (Jones *et al.*, 1990). Analysis of the chicken TN promoter region revealed a number of potential binding sites for homeodomain proteins of the Antennapedia and Engrailed classes which contained the TAAT sequence (Desplan *et al.*, 1988) and the octamer motif ATGCAAAT, which is known to bind POU-homeodomain proteins (Scheidreit *et al.*, 1988; Scholer *et al.*, 1988).

HOMEOBOX TRANSCRIPTION FACTORS AND THE CONTROL OF TENASCIN EXPRESSION

Based on the dynamic expression patterns of tenascin described above, an attractive hypothesis was that homeodomain proteins encoded by different homeobox genes control the spatiotemporal expression of TN during embryogenesis (Jones *et al.*, 1990; Edelman, 1992; Edelman and Jones, 1992). To provide evidence for this hypothesis, the TN promoter was tested for binding and regulation by the products of specific homeobox genes. Various regions containing the putative homeodomain binding sites were assayed using DNAse I footprinting for their ability to bind such homeodomain proteins (Jones *et al.*, 1992). While many of the TAAT motifs were examined, one region in particular was found to bind an Antennapedia class homeodomain protein encoded by the *Drosophila* segmentation gene, *fushi tarazu*. To determine whether homeodomain proteins could actually regulate TN gene expression, a series of co-transfection experiments was conducted in NIH3T3 cells. Various gene regulatory constructs containing different segments of the chicken TN promoter linked to a bacterial chloramphenicol acetyltransferase (CAT) reporter were co-transfected together with plasmids driving the expression of homeobox genes (Jones *et al.*, 1992). Co-transfection with a plasmid expressing *evx-1* (a mouse homeobox- containing gene most similar to the *eve* gene from *Drosophila*) stimulated TN promoter activity whereas cotransfection in control experiments with a plasmid expressing *hoxA5* (*hox1.3*) had no effect. An 89 base pair segment of the TN promoter was found to be responsible for activation by *evx-1*. It contained a TRE/AP-1 element and mutation of this DNA sequence abrogated the response (Jones *et al.*, 1992). The TRE/AP-1 element is known to bind transcription factors encoded by the *fos* and *jun* gene family. These factors are targets of signal transduction pathways that are activated by growth factors (Angel *et al.*, 1987; Curran and Franza, 1988). These findings raise the possibility that regulation of TN promoter activity by expression of *evx-1* may be mediated through Fos or Jun proteins which bind at the TRE/AP-1 site. In separate transfection experiments, it has been shown that the TRE/AP-1 is also required for TN promoter response to fetal bovine serum (Jones *et al.*, 1992). Interestingly, deletion of the TN promoter region that contained the TAAT motif which was found to bind the *fushi tarazu* homeodomain protein reduced the level of activation by *evx-1* but did not abrogate this response completely. These results indicated that the TRE/AP-1 sequence was a critical component for the activation of the TN promoter by *evx-1* but additional sequences such as the TAAT motif were required for full induction. These observations do not necessarily imply that the product of *evx-1* is a natural modulator of TN promoter activity *in vivo*. Indeed, using the same cotransfection system, we found that TN promoter activity can be controlled by a number of other homeobox genes and these effects can vary depending on the cell line used. We have found, for example, that expression of *hoxA1*, *hoxB9*, *hoxD9*, *pax-3*, and *pax-6*, which encode many different classes of homeodomain proteins activate the TN promoter (FSJ, DWC, and G.M. Edelman, unpublished data). It thus remains a challenge to determine whether homeodomain proteins control specific contexts of TN expression *in vivo*.

Recently a TN-like gene from *Drosophila*, called *ten^{m}/odd Oz* has been identified (Baumgartner *et al.*, 1994; Levine *et al.*, 1994). This gene encodes a protein with EGF-like repeats and FN type III domains. Interestingly, *ten^{m}/*

odd Oz is a pair-rule segmentation gene expressed in seven stripes that overlap with the *even-skipped* stripes and its expression is controlled by *even-skipped* and *fushi tarazu* gene products (Baumgartner *et al.*, 1994). These observations suggest that the developmental expression patterns of the extended family of TN-like molecules may be controlled by the product of homeobox containing genes in many different and widely divergent species. It would be of interest to isolate the promoter region from the *ten^m/odd Oz* gene and determine whether it, too, is a target of homeobox-containing genes.

Observations on the chicken TN promoter and its activation by homeobox-containing genes prompt the identification and characterization of similar regulatory sequences for the TN gene in other vertebrate species. It will be of particular interest to compare the important DNA regulatory elements and mechanisms used in the control of the expression of genes that encode TN-like molecules, such as TN-R (also known as $J1_{160/180}$ or restrictin), and TN-X. Such comparisons may uncover an evolutionarily conserved mechanism for developmental control of the TN family by homeobox genes. These studies may also yield clues about whether other members of the TN family can compensate for the loss of a particular TN gene in knockout experiments.

TENASCIN REGULATORY SEQUENCES FROM OTHER SPECIES AND CONSERVATION OF ELEMENTS

Recently a 4 kb sequence for the mouse tenascin promoter has been determined in an effort to determine whether common or different DNA regulatory elements control the expression of this gene in different species (Copertino *et al.*, 1995). In addition, 2 kb of the upstream regulatory regions for the human TN gene have been characterized (Gherzi *et al.*, 1995). Comparison of the TN promoter from all three species (chicken, mouse, and human) illustrate commonalities and differences in the regulatory strategies of TN in all these species. The proximal regions of the mouse and chicken promoters show remarkable structural and functional similarities suggesting that the regulatory mechanism controlling expression of the TN gene may have been conserved during evolution (Copertino *et al.*, 1995). Both the chicken and mouse TN genes show a single RNA start site for TN mRNA. The DNA sequences upstream of this start site in both the chicken and mouse TN genes, show significant similarities in sequence. The sequence similarity extended from the 5′ end of the mRNAs for the chicken and mouse to approximately 230 base pairs upstream of the transcriptional initiation site. The DNA sequence identity within this region was 79% and could be divided into five major regions, which included two separate dA-dT sequences, the octamer motif, and a TAAT motif that resembled a binding site for homeodomain proteins and the site of assembly of the basic transcription machinery including the TATA box and the cap site (Table 1.2). This proximal promoter region is responsible for most of the positive regulatory activity for both the chicken and mouse TN promoters. The mouse promoter also conserved the TRE/AP-1 sequence. However, this element was located in a different position in the mouse gene further upstream from the corresponding position of 279 in the chicken gene.

Comparison of these TN promoters also showed several differences in the regulatory sequences (see Table 1.2). There are both temporal and spatial

Table 1.2 Putative DNA regulatory elements shared among tenascin promoters

Element	*Species*	*Position*	*Sequence*
TATA box	human	–26 to –19	TATATAAG
	mouse	–35 to –28	TATATAAG
	chicken	–28 to –21	TATATAAA
octamer	human	–186 to –179	ATGCAAAT
	mouse	–201 to –194	ATGCAAAT
	chicken	–204 to –197	ATGCAAAT
NF-1	human	–863 to –851	TGGAGGAAATCCA
		–166 to –154	TGGCCGAGAGCCA
	mouse	–182 to –170	TGGCGGCGCGCCT
	chicken	–63 to –52	TGG-AATGAGCCA
AP-2	human	387 to –380	GGCGAGCCC
	mouse	–333 to –326	CCCAAGGC
	chicken	?	
HD	human	–50 to –41	CCCATTACAG
	mouse	–59 to –52	CCCATTACAG
	chicken	–73 to –64	CCTATTACAG
Krox-20	human	–230 to –222	TCGGGGGCG
	mouse	–256 to –248	GCGGGGGCG
	chicken	NP	
SP1	human	–227 to –218	GGGGCGGGCC
	mouse	NP	
	chicken	–441 to –447	GGGGCGG
T-rich	human	NP	
	mouse	–230 to –214	TACTTTTTTTTTTCTTC
	chicken	–259 to –243	TTTTTTTTTTTTTGTTT
A-rich	human	–86 to –78	TG- - -AAAAA- - - -GT
	mouse	–102 to –87	TGAAAAAAAAAAAAGT
	chicken	–98 to –83	TG-AAAAAAAAAA-TG
TRE/AP-1	human	–865 to –869	TGACTCA
	mouse	–1464 to –1456	TGAATCA
		–1433 to –1425	TGAGTCG
	chicken	–280 to –274	TGAGTCA

NP — not present, ? — presence unknown

differences in the neural expression patterns of TN among vertebrate species. Such differences may result from the diversification of regulatory elements used in the species-specific patterning of TN against a relatively constant role of the proximal promoter region. These notions will have to be tested in a variety of *in vitro* cellular and transgenic animal experiments.

Recent data from the human TN promoter shows that it is approximately 96% identical in the proximal promoter region to that of the mouse. Identities include the elements that are conserved with the chicken but in addition include elements such as the Krox-20 binding site. Tenascin and Krox-20 may be a part of the same evolutionarily conserved regulatory network that directs segmentation. Krox-20 is restricted to two odd-numbered rhombomeres (3 and 5) in the developing mouse hindbrain and knockout of the Krox-20 gene in mice shows a reduction or elimination of these structures (Schneider-Maunoury *et al.*, 1993). These effects are similar to segmentation defects observed in *Drosophila* when the function or expression of primary pair-rule genes is either disrupted or eliminated (Schneider-Maunoury *et al.*, 1993). The expression of the TN-like gene *Drosophila ten*^{m}*/odd Oz* in seven stripes (parasegments) is characteristic of a secondary pair-rule gene. These data support the notion that TN in vertebrates and similar molecules in *Drosophila* may be targets of the Krox family of segmentation genes. The Krox GC element which is highly conserved between the mouse and human TN promoters was not found in the chicken TN promoter. It is possible that the GC-rich sequence resembling an SP1 motif in the chicken TN promoter may be a target of Krox transcription factors. Alternatively, the expression of the chicken TN gene may not be controlled by Krox proteins. Thus, the absence of this regulatory element would represent a species-specific difference in the control of TN gene expression.

The TN promoters from the chicken, mouse and human contain a sequence that resembles a binding site for nuclear factor 1 (NF-1). It has been shown in previous studies that the NF-1 binding site in a collagen gene controls responsivity to signals from the TGF-β growth factor family (Rossi *et al.*, 1988). Thus, it will be interesting to determine whether the NF-1 sites in the TN genes are responsible for induction by growth factors of the TGF-β family. In recent experiments, we have determined that in NIH3T3 cells, the mouse TN promoter is highly active, whereas the chicken is not. In DNAse I footprinting studies, we found that the NF-1 sequence in the mouse promoter is protected from DNase digestion using nuclear extracts from NIH3T3 cells (FSJ, DWC, and G.M. Edelman, unpublished results). In addition, deletion of the NF-1 site diminishes significantly the positive activity of TN in NIH3T3 cells. These data suggest that transcription factors of the NF-1 family may be involved in the constitutive and possibly the inducible expression of TN in the mouse and prompts a similar examination of the involvement of NF-1 in the regulation of the TN gene in other species.

Although preliminary, recent studies suggest that both the chicken and mouse TN promoters respond to *pax* gene products (FSJ, DWC, and G.M. Edelman, unpublished data). Pax proteins are important regulators of neural differentiation (Chalepakis *et al.*, 1993). Studies are well underway to identify the regulatory sequences that are important for regulation of TN in the nervous system by different *pax* gene products. Analysis of the human TN promoter has also located a number of putative binding sites for pax proteins (Gherzi *et al.*, 1995). Glucocorticoids have also been found to downregulate expression of mouse TN (Ekblom *et al.*, 1993). A consensus site for the binding of glucocorticoid receptors has been located in the chicken TN promoter sequence at -985. The control of TN expression by growth factors and glucocorticoids will be an important factor in understanding the role of TN in tumor formation and metastasis.

Recently, Chiquet-Ehrismann has used the chicken tenascin promoter to study the effects of cellular mechanical stress on TN gene expression. Tenascin expression was found to increase in fibroblasts in culture when they were exposed to mechanical stress. The region required for this response was localized to a 100 base pair segment of the chicken TN promoter which does not contain any elements for known transcription factors (Chiquet-Ehrismann *et al.*, 1994). These studies show that TN gene expression is responsive to changes in the tensile forces that arise from alterations in cell morphology. Regulation of gene expression by mechanical stress may be important during embryogenesis and other tissue remodelling processes such as wound healing.

CONCLUSION

The study of tenascin, its splice variants, the regulation of its expression provides an exciting field of investigation for the future. Since the cell binding properties of the fibronectin type III repeats have been well established (see Chapter 2), it may be time to expand functional analyses of TN to domains which have received little attention. For instance, the functions of the fibrinogen-like distal domain sequences from TN and other potentially interesting molecules such as PT49 (Koyama *et al.*, 1987) and scabrous (Baker *et al.*, 1990) are yet to be discovered. An equal challenge in TN biology will be to ascribe specific functions to the different splice variants that are found for TN during development and in disease states. Finally, understanding the regulatory mechanisms that control the different contexts of TN gene expression will be a formidable challenge. The use of particular DNA control elements and *trans*-acting factors in the regulation of TN gene expression may also extend to the regulatory programs of other TN-related molecules.

REFERENCES

1. Angel, P., Imagawa, M., Chiu, R., Stein, B., Imbra, R.J., Rahmsdorf, H.J., Jonat, C., Herrlich, P., and Karin, M. (1987) Phorbol ester-inducible genes contain a common *cis* element recognized by a TPA-modulated *trans*-acting factor. *Cell*, **49**, 729–739.
2. Aufderheide, E., and Ekblom, P. (1988) Tenascin during gut development: appearance in the mesenchyme, shift in molecular forms, and dependence on epithelial-mesenchymal interactions. *J. Cell Biol.*, **105**, 2341–2349.
3. Bacon-Baguley, T., Kudryk, B.J., and Walz, D.A. (1987) Thrombospondin interaction with fibrinogen. *J. Biol. Chem.*, **262**, 1927–1930.
4. Baker, N.E., Mlodzik, M., and Rubin, G.M. (1990) Spacing differentiation in the developing *Drosophila* eye: A fibrinogen-related lateral inhibitor encoded by *scabrous*. *Science*, **254**, 1370–1376.
5. Baron, M., Main, A.L., Driscoll, P.C., Mardon, H.J., Boyd, J., and Campbell, I.A. (1992) ^{3}H-NMR assignment and secondary structure of the cell adhesion type III module of fibronectin. *Biochemistry*, **31**, 2068–2073.
6. Bartsch, S., Bartsch, U., Dories, U., Faissner, A., Weller, A., Ekblom, P., and Schachner, M. (1992) Expression of tenascin in the developing cerebellar cortex. *J. Neurosci.*, **12**, 736–749.
7. Baumgartner, S., Martin, D., Hagios, C., and Chiquet-Ehrismann, R. (1994) *ten*^{m}, a *Drosophila* gene related to tenascin, is a new pair-rule gene. *EMBO J.*, **13**, 3728–3740.
8. Borsi, L., Balza, E., Castellani, P., Carnemolla, B., Ponassi, M., Querzé, G., and Zardi, L. (1994) Cell-cycle dependent alternative splicing of the tenascin primary transcript. *Cell Adh. Comm.*, **1**, 307–317.
9. Borsi, L., Balza, E., Gaggero, B. Allemanni, G., and Zardi, L. (1995) The alternative splicing pattern of the tenascin-C pre-mRNA is controlled by the extracellular pH. *J. Biol. Chem.*, **270**, 6243–6245.
10. Borsi, L., Carnemolla, B., Nicclo, G., Spina, B., Tanara, G., and Zardi, L. (1992) Expression of different tenascin isoforms in normal hyperplastic and neoplastic human breast tissues. *Int. J. Cancer*, **52**, 688–692.
11. Bourdon, M.A., Wikstrand, C.J., Furthmayer, H., Matthews, T.J., and Bigner, D.D. (1983) Human glioma-mesenchymal extracellular matrix antigen defined by monoclonal antibody. *Cancer Res.*, **43**, 2796–2805.
12. Bristow, J., Tee, M.K., Gitelman, S.E., Mellon, S.H., and Miller, W.L. (1993) Tenascin-X: A novel extracellular matrix protein encoded by the human XB gene overlapping P450c21B. *J. Cell Biol.*, **122**, 265–278.
13. Chalepakis, G., Stoykova, A., Wijnholds, J., Tremblay, P., and Gruss, P. (1993) Pax: gene regulators in the developing nervous system. *J. Neurobiol.*, **24**, 1367–1384.
14. Chiquet, M., and Fambrough, D.M. (1984) Chick myotendinous antigen. I. A monoclonal antibody as a marker for tendon and muscle morphogenesis. *J. Cell Biol.*, **98**, 1926–1936.
15. Chiquet-Ehrismann, R., Mackie, E.J., Pearson, C.A., and Sakakura, T. (1986) Tenascin: an extracellular matrix protein involved in tissue interactions during fetal development and oncogenesis. *Cell*, **47**, 131–139.

16. Chiquet-Ehrismann, R., Matsuoka, Y., Hofer, U., Spring, J., Bernasconi, C., and Chiquet, M. (1991) Tenascin variants: differential binding to fibronectin and distinct distribution in cell cultures and tissues. *Cell Reg.*, **2**, 927–938.
17. Chiquet-Ehrismann, R., Tannheimer, M., Koch, M., Brunner, A., Spring, J., Martin, D., Baumgartner, S., and Chiquet, M. (1994) Tenascin-C expression by fibroblasts is elevated in stressed collagen gels. *J. Cell Biol.*, **127**, 2093–2101.
18. Chung, C.Y., and Erickson, H.P. (1994) Cell surface annexin II is a high affinity receptor for the alternatively spliced segment of Tenascin-C. *J. Cell Biol.*, **126**, 539–548.
19. Chuong, C.-M., and Chen, H.-M. (1991) Enhanced expression of neural cell adhesion molecules and tenascin (cytotactin) during wound healing. *Am. J. Pathol.*, **138**, 427–440.
20. Conway, J.F., and Parry, D.A.D. (1991) Three-stranded α-fibrous proteins: the heptad repeat and its implications for structure. *Int. J. Biol. Macromol.*, **13**, 14–16.
21. Cooke, R.M., Wilkinson, A.J., Baron, M., Pastore, A., Tappin, M.J., Campbell, I.D., Gregory, H., and Sheard, B. (1987) The solution structure of human epidermal growth factor. *Nature (London)*, **327**, 339–341.
22. Copertino, D.W., Jenkinson, S., Jones, F.S., and Edelman, G.M. (1995) Structural and functional similarities between the promoters for mouse tenascin and chicken cytotactin. *Proc. Natl. Acad. Sci. USA*, **92**, 2131–2135.
23. Crossin, K.L. (1994) The functional role of cytotactin/tenascin in morphogenesis: a modest proposal. *Persp. Dev. Neurobiol.*, **2**, 21–32.
24. Crossin, K.L., Hoffman, S., Grumet, M., Thiery, J.-P., and Edelman, G.M. (1986) Site-restricted expression of cytotactin during development of the chicken embryo. *J. Cell Biol.*, **102**, 1917–1930.
25. Curran, T., and Franza, B.R. (1988) Fos and Jun: The Ap-1 connection. *Cell*, **55**, 395–397.
26. Desplan, C., Theis, J., and O'Farrell, P.H. (1988) The sequence specificity of homeodomain-DNA interaction. *Cell*, **54**, 1081–1090.
27. Doolittle, R.F. (1984) Fibrinogen and fibrin. *Annu. Rev. Biochem.*, **53**, 195–229.
28. Dörries, U., and Schachner, M. (1994) Tenascin mRNA isoforms in the developing mouse brain. *J. Neurosci. Res.*, **37**, 336–347.
29. Edelman, G.M. (1992) Morphoregulation. *Dev. Dynamics*, **193**, 2–10.
30. Edelman, G.M., and Jones, F.S. (1992) Cytotactin: a morphoregulatory molecule and a target for regulation by homeobox gene products. *Trends Biochem. Sci.*, **17**, 228–232.
31. Ekblom, M., Fässler, R., Tomasini-Johansson, B., Nilsson, K., and Ekblom, P. (1993) Downregulation of tenascin expression by glucocorticoids in bone marrow stromal cells and in fibroblasts. *J. Cell Biol.*, **123**, 1037–1045.
32. Erickson, H.P., and Bourdon, M.A. (1989) Tenascin: an extracellular matrix protein prominent in specialized embryonic tissues and tumors. *Annu. Rev. Cell Biol.*, **5**, 71–92.
33. Erickson, H.P., and Iglesias, J.L. (1984) A six-armed oligomer isolated from cell surface fibronectin preparations. *Nature (London)*, **311**, 267–269.
34. Faissner, A., and Kruse, J. (1990) J1/tenascin is a repulsive substrate for central nervous system neurons. *Neuron*, **5**, 627–637.

35. Friedlander, D.R., Hoffman, S., and Edelman, G.M. (1988) Functional mapping of cytotactin: proteolytic fragments active in cell-substrate adhesion. *J. Cell Biol.*, **107**, 2329–2340.
36. Fuss, B., Wintergerst, E.-S., Bartsch, U., and Schachner, M. (1993) Molecular characterization and in situ mRNA localization of the neural recognition molecule J1-160/180: a modular structure similar to tenascin. *J. Cell Biol.*, **120**, 1237–1249.
37. Gherzi, R., Carnemolla, B., Siri, A., Ponassi, M., Balza, E., and Zardi, L. (1995) Human tenascin gene: Structure of the 5′-region, identification, and characterization of the transcription regulatory sequences. *J. Biol. Chem.*, **7**, 1–6.
38. Grumet, M., Hoffman, S., Crossin, K.L., and Edelman, G.M. (1985) Cytotactin, an extracellular matrix protein of neural and non-neural tissues that mediates glia-neuron interaction. *Proc. Natl. Acad. Sci. USA*, **82**, 8075–8079.
39. Gulcher, J.R., Nies, D.E., Marton, L.S., and Stefansson, K. (1989) An alternatively spliced region of the human hexabrachion contains a novel repeat of potential N-glycosylation sites. *Proc. Natl. Acad. Sci. USA*, **86**, 1588–1592.
40. Handford, P.A., Baron, M., Mayhew, M., Willis, A., Beesley, T., Brownlee, G.G., and D, C.I. (1990) The first EGF domain from human factor IX contains a high affinity calcium binding site. *EMBO J.*, **9**, 475–480.
41. Hoffman, S., Crossin, K.L., and Edelman, G.M. (1988) Molecular forms, binding functions, and developmental expression patterns of cytotactin and cytotactin-binding proteoglycan, an interactive pair of extracellular matrix molecules. *J. Cell Biol.*, **106**, 519–532.
42. Humbert-David, N., and Garrone, R. (1993) A six-armed, tenascin-like protein extracted from the Porifera *Oscarella tuberculata* (Homosclerophorida). *Eur. J. Biochem.*, **216**, 255–260.
43. Inaguma, Y., Kusakabe, M., Mackie, E.J., Pearson, C.A., Chiquet-Ehrismann, R., and Sakakura, T. (1988) Epithelial induction of stromal tenascin in the mouse mammary gland: from embryogenesis to carcinogenesis. *Dev. Biol.*, **128**, 245–255.
44. Jones, F.S. (1991) The molecular biology of the extracellular matrix glycoprotein cytotactin: Primary structure, alternative splicing, and characterization of the promoter. In: *Ph.D. Thesis, The Rockefeller University*, University of Michigan Press, Ann Arbor.
45. Jones, F.S., Burgoon, M.P., Hoffman, S., Crossin, K.L., Cunningham, B.A., and Edelman, G.M. (1988) A cDNA clone for cytotactin contains sequences similar to epidermal growth factor-like repeats and segments of fibronectin and fibrinogen. *Proc. Natl. Acad. Sci. USA*, **85**, 2186–2190.
46. Jones, F.S., Chalepakis, G., Gruss, P., and Edelman, G.M. (1992) Activation of the cytotactin promoter by the homeobox-containing gene, Evx-1. *Proc. Natl. Acad. Sci. USA*, **89**, 2091–2095.
47. Jones, F.S., Crossin, K.L., Cunningham, B.A., and Edelman, G.M. (1990) Identification and characterization of the promoter for the cytotactin gene. *Proc. Natl. Acad. Sci. USA*, **87**, 6497–6501.
48. Jones, F.S., Hoffman, S., Cunningham, B.A., and Edelman, G.M. (1989) A detailed structural model of cytotactin: Protein homologies, alternative RNA splicing, and binding regions. *Proc. Natl. Acad. Sci. USA*, **86**, 1905–1909.

49. Kaplony, A., Zimmermann, D.R., Fischer, R.W., Imhof, B.A., Odermatt, B.F., Winterhalter, K.H., and Vaughan, L. (1991) Tenascin Mr 220,000 isoform expression correlates with corneal cell migration. *Development*, **112**, 605–614.
50. Koyama, T., Hall, L.R., Haser, W.G., Tonegawa, S., and Saito, H. (1987) Structure of a cytotoxic T-lymphocyte-specific gene shows a strong homology to fibrinogen beta and gamma chains. *Proc. Natl. Acad. Sci. USA*, **84**, 1609–1613.
51. Kruse, J., Keilhauer, G., Faissner, A., Timpl, R., and Schachner, M. (1985) The J1 glycoprotein- A novel nervous system cell adhesion molecule of the L2/HNK-1 family. *Nature (London)*, **316**, 146–148.
52. Kruse, J., Mailhammer, R., Wernecke, H., Faissner, A., Sommer, I., Goridis, C., and Schachner, M. (1984) Neural cell adhesion molecules and myelin-associated glycoprotein share a common carbohydrate moiety recognized by monoclonal antibodies L2 and HNK-1. *Nature (London)*, **311**, 151–153.
53. Leahy, D.J., Hendrickson, W.A., Aukhil, I., and Erickson, H.P. (1992) Structure of a fibronectin type III domain from tenascin phased by MAD analysis of the selenomethionyl protein. *Science*, **258**, 987–991.
54. Levine, A., Bashan-Ahrend, A., Budai-Hadrian, O., Gartenberg, D., Menasherow, S., and Wides, R. (1994) *odd Oz*: A novel *Drosophila* pair-rule gene. *Cell*, **77**, 587–598.
55. McGinnis, W., and Krumlauf, R. (1992) Homeobox genes and axial patterning. *Cell*, **68**, 283–302.
56. McKeown, M. (1992) Alternative mRNA splicing. *Annu. Rev. Cell Biol.*, **8**, 133–155.
57. Morel, Y., Bristow, J., Gitelman, S.E., and Miller, W.L. (1989) Transcript encoded on the opposite strand of the human steroid 21-hydroxylase/complement component C4 gene locus. *Proc. Natl. Acad. Sci. USA*, **86**, 6582–6586.
58. Nishi, T., Weinstein, J., Gillespie, W.M., and Paulson, J.C. (1991) Complete primary structure of porcine tenascin. Detection of tenascin transcripts in adult submaxillary glands. *Eur. J. Biochem.*, **202**, 643–648.
59. Norenberg, U., Wille, H., Wolff, J.M., Frank, R., and Rathjen, F.G. (1992) The chicken neural extracellular matrix molecule restrictin: similarity with EGF-, fibronectin type III-, and fibrinogen-like motifs. *Neuron*, **8**, 849–863.
60. Patthy, L. (1990) Homology of a domain of the growth hormone/prolactin receptor family with type III modules of fibronectin. *Cell*, **61**, 13–14.
61. Pearson, C.A., Pearson, D., Shibahara, S., Hofsteenge, J., and Chiquet-Ehrismann, R. (1988) Tenascin: cDNA cloning and induction by TGF-β. *EMBO J.*, **7**, 2977–2982.
62. Prieto, A.L., Andersson-Fisone, C., and Crossin, K.L. (1992) Characterization of multiple adhesive and counteradhesive domains in the extracellular matrix protein cytotactin. *J. Cell. Biol.*, **119**, 663–678.
63. Prieto, A.L., Jones, F.S., Cunningham, B.A., Crossin, K.L., and Edelman, G.M. (1990) Localization during development of alternatively-spliced forms of cytotactin by *in situ* hybridization. *J. Cell Biol.*, **111**, 685–698.
64. Probstmeier, R., Kuhn, K., and Schachner, M. (1990) Fibroblast-derived J1 adhesion glycoproteins show binding properties to extracellular matrix constituents different from those of central nervous system. *J. Neurochem.*, **54**, 1016–1019.

65. Rettig, W.J., Triche, T.J., and Garin-Chesa, P. (1989) Stimulation of human neuronectin secretion by brain-derived growth factors. *Brain Res.*, **487**, 171–177.
66. Rossi, P., Karsenty, G., Roberts, A.B., Roche, N.S., Sporn, M.B., and de Combrugghe, B. (1988) A nuclear factor 1 binding site mediates the transcriptional activation of a type I collagen promoter by transforming growth factor-β. *Cell*, **52**, 405–414.
67. Scheidreit, C., Cromlish, J.A., Gerster, T., Kauakami, K., Balmaceda, C.-G., Currie, R.A., and Roeder, R.G. (1988) A human lymphoid-specific transcription factor that activates immunoglobulin genes is a homeobox protein. *Science*, **336**, 551–557.
68. Schneider-Maunoury, S., Topilko, P., Seitanidou, T., Levi, G., Cohen-Tannoudji, M., Pournin, S., Babinet, C., and Charnay, P. (1993) Disruption of *Krox-20* results in alteration of rhombomeres 3 and 5 in the developing hindbrain. *Cell*, **75**, 1199–1214.
69. Scholer, H.R., Hatzopoulos, A.K., Balling, R., Suzuki, N., and Gruss, P. (1988) A family of octamer-specific proteins present during mouse embryogenesis: evidence for germline-specific expression of an Oct factor. *EMBO J.*, **3**, 2543–2550.
70. Siri, A., Carnemolla, B., Saginati, M., Leprini, A., Casari, G., Baralle, F., and Zardi, L. (1991) Human tenascin: primary structure, pre-mRNA splicing patterns and localization of the epitopes recognized by two monoclonal antibodies. *Nucl. Acids Res.*, **19**, 525–531.
71. Smith, C.W.J., Patton, J.G., and Nadal-Ginard, B. (1989) Alternative splicing in the control of gene expression. *Annu. Rev. Genet.*, **23**, 527–577.
72. Spring, J., Beck, K., and Chiquet-Ehrismann, R. (1989) Two contrary functions of tenascin: dissection of the active sites by recombinant tenascin fragments. *Cell*, **59**, 325–334.
73. Sriramarao, P., Mendler, M., and Bourdon, M.A. (1993) Endothelial cell attachment and spreading on human tenascin is mediated by $\alpha 2\beta 1$ and $\alpha v\beta 3$ integrins. *J. Cell Sci.*, **105**, 1001–1012.
74. Tan, S.-S., Crossin, K.L., Hoffman, S., and Edelman, G.M. (1987) Asymmetric expression in somites of cytotactin and its proteoglycan ligand is correlated with neural crest cell distribution. *Proc. Natl. Acad. Sci. USA*, **84**, 7977–7981.
75. Tucker, R., and McKay, S.E. (1991) The expression of tenascin by neural crest cells and glia. *Development*, **112**, 1031–1039.
76. Tuffy, R., and Kretsinger, R.H. (1975) Troponin and paralbumin calcium binding regions predicted in myosin light chain and lysozyme. *Science*, **187**, 167–169.
77. Weller, A., Beck, S., and Ekblom, P. (1991) Amino acid sequence of mouse tenascin and differential expression of two tenascin isoforms during embryogenesis. *J. Cell Biol.*, **112**, 355–362.
78. Wright, D.W., and Mayne, R. (1988) Vitreous humor of chicken contains two fibrillar systems: an analysis of their structure. *J. Ultramol. Str. Res.*, **100**, 224–234.
79. Xu, X., and Doolittle, R.F. (1990) Presence of a vertebrate fibrinogen-like sequence in an echinoderm. *Proc. Natl. Acad. Sci. USA*, **87**, 2097–2101.

2. THE ROLE OF TENASCIN DOMAINS IN DEVELOPMENTAL PATTERNING, CELL MIGRATION AND GENE EXPRESSION

KATHRYN L. CROSSIN*, ANNE L. PRIETO and VINCENT P. MAURO

Department of Neurobiology, The Scripps Research Institute
10666, North Torrey Pines Road, La Jolla, CA 92037

INTRODUCTION

The tenascin (TN) protein appears to contribute to important functions of embryonic extracellular matrices and may play a role in the regulation of cell migration, process extension, and cell adhesion to other basement membrane proteins. Its expression pattern is highly dynamic during development and it is reexpressed in a large number of tumors and in wounds during the inflammatory process as discussed in detail in Chapter 4. Its dynamic distribution pattern during development and in disease states, as well as its multidomain structure emphasized in Chapter 1 suggest that a dissection of its structure-function relationships will be revealing in understanding how the protein contributes to various cellular processes during development.

While the multidomain structure of TN is well-described in Chapter 1, it is worth reviewing briefly here since various proteolytic and recombinant fragments comprising subdomains of the protein as well as domain-specific antibodies have been used to establish structure-function relationships. The domain structure of TN is reviewed in Figure 2.1 which includes reference to the mapping of recombinant protein fragments and monoclonal antibodies to be discussed below. The TN molecule consists of four major regions. At the amino terminus is a cysteine-rich region which is presumably responsible for the disulfide bonding of monomers into multimers such as the hexabrachion structure. This region is homologous to the amino terminal domains of restrictin (TN-R) (Rathjen *et al.*, 1991; Fuss *et al.*, 1993) and TN-X (Matsumoto *et al.*, 1992; Bristow *et al.*, 1993), the TN-related molecules that have thus far been described. The cysteine rich region is followed by a series of epidermal growth factor (EGF) -like repeats of 13½ or 14½ domains depending on the animal species of origin. This series is followed by a number of repeats homologous to the type III repeats in fibronectin, the number of which varies as a result of alternative splicing of the RNA, yielding a number of different isoforms of the TN protein *in vivo*. The smallest TN protein isoform contains 8 fibronectin

* Corresponding author

type III repeat homology regions In the chicken molecule, the major isoforms contain one or three additional fibronectin type III repeats; the amounts of each isoform differ among individual tissues (Aufderheide and Ekblom, 1988; Prieto *et al.*, 1990; Tucker and McKay, 1991). Additional FN type III domains are found in rare isoforms in the chicken and in a variety of mammalian species (see Chapter 1). The carboxyterminus is homologous to the β and γ chains of fibrinogen and contains a potential calcium binding domain (Jones *et al.*, 1990).

TN can mediate cell-substratum adhesion, but it also exhibits counteradhesive properties that depend upon the presence of particular cell surface receptors as well as on interactions with other extracellular matrix molecules. The protein thus has amphitropic effects *in vivo*. The combination of seemingly antagonistic functions and highly regulated, place-dependent expression patterns may provide powerful constraints on developing morphology in the embryo, particularly in the nervous system (see Chapter 3). It may also affect cellular responses in a variety of disease states (see Chapter 4).

PLACE-DEPENDENT EXPRESSION PATTERNS OF TN DURING DEVELOPMENT

One of the striking features of TN that prompted extensive interest in the molecule was its site-restricted distribution throughout embryonic development. The molecule is expressed in an unusual cephalocaudal gradient at the time of gastrulation, neurulation, and during somite formation in the chick (Crossin *et al.*, 1986). This cephalocaudal pattern may be related to the ability of the TN gene to be controlled *in vivo* by the products of homeobox- and paired box-containing genes and by growth factors, as suggested by *in vitro* studies (see Chapter 1).

An example of the patterned and dynamic distribution of TN occurs during the formation and differentiation of somites both prior to and during the migration of neural crest cells. Somites are segmental blocks of mesoderm in vertebrates that give rise to muscles, vertebral cartilage and bones, and impose a segmental structure on the peripheral nervous system by limiting the migration of neural crest cells and the outgrowth of motor axons to the anterior part of each somite. Molecules present in the somites, at the cell surface, and in the extracellular matrix, may therefore play a crucial role in these processes. In the epithelial somites, the TN protein and mRNA (Crossin *et al.*, 1986; Tan *et al.*, 1987; Stern *et al.*, 1989; Tan *et al.*, 1991) are expressed only in the caudal half-somite. Later, as the somite becomes mesenchymal, neural crest cells and motor axons enter only the rostral half of the sclerotome; at this time, fibronectin is present uniformly throughout the sclerotome (Duband *et al.*, 1987), but the TN protein and mRNA (Tan *et al.*, 1987; Mackie *et al.*, 1988a; Tan *et al.*, 1991) become restricted to the rostral half-sclerotome

(Tan *et al.*, 1987). The distribution of TN in the sclerotome suggests that this molecule may affect patterning in the peripheral nervous system.

Neural crest extirpation experiments (Tan *et al.*, 1991) demonstrated that somite cells synthesize high molecular weight TN in the absence of neural crest cells, and thus neural crest cells do not appear to regulate the expression of TN by the sclerotomal cells as previously proposed (Stern *et al.*, 1989). The studies of Tan *et al.*, (1991) also showed that the presence of TN is neither necessary nor sufficient for the appropriate patterning of the neural crest cells, although in a number of experimental cases, the pattern of TN in the somite did not change when neural crest cells were removed (Tan *et al.*, 1987; Tan *et al.*, 1991). In other cases, however, the pattern of TN expression was changed simply by mock extirpation in which the crest cells were left *in situ* and migrated normally (Tan *et al.*, 1991). These findings suggest that the pattern of appearance of TN after neural crest cell removal was in part a response to the wounding process. A number of studies in other systems have indicated a role for TN in modulating wound healing (Mackie *et al.*, 1988b; Lightner *et al.*, 1989) during which marked increases in TN expression were observed.

These observations do not preclude a role for TN in neural crest migration *in vivo*. For example, cranial neural crest migration was perturbed upon introduction of TN antibodies *in vivo* (Bronner-Fraser, 1988), supporting the hypothesis that TN plays a role in neural crest migration in this region. In *in vitro* chick neural crest cell migration assays, fibronectin coated onto a two-dimensional substrate promotes neural crest cell migration (Rovasio *et al.*, 1983). When TN was combined with fibronectin on the substrate, the amount of migration on the fibronectin substrate was decreased (Tan *et al.*, 1987). In cultures of quail neural crest cells migrating on native basal lamina substrates, the addition of TN to the culture medium reduced the rate of cell migration (Halfter *et al.*, 1989). Neural crest cells on TN substrates alone or on TN mixed with fibronectin exhibited a much less spread morphology (Tan *et al.*, 1987; Epperlein *et al.*, 1988; Mackie *et al.*, 1988a; Halfter *et al.*, 1989). Similarly, a TN binding proteoglycan (originally termed the CTB proteoglycan (Hoffman and Edelman, 1987; Hoffman *et al.*, 1988)), when combined with fibronectin, also reduced cell spreading and migration on fibronectin (Tan *et al.*, 1987). The TN-binding proteoglycan may inhibit the entry of neural crest cells into the caudal half of the sclerotome where this proteoglycan is normally expressed *in vivo* (Tan *et al.*, 1987). It is possible that TN synthesized *in vivo* by the rostral somite or in the cranial mesenchyme acts as a substrate that blocks the further migration of neural crest cells, facilitating their aggregation via various cell adhesion molecules (Duband *et al.*, 1987) and subsequent ganglion formation.

During the development of the central and peripheral nervous systems, TN has been variously implicated in numerous cellular processes including neuronal migration, axonal guidance, alteration of dendritic morphology, synaptogenesis, and myelination. The role of TN in these processes has been deduced from localization studies of TN mRNA and protein and through antibody perturbation experiments. For example, during formation of the visual system, TN is seen in areas bordering projecting retinal ganglion axons including the optic disc and the outer surface of the optic nerve (Perez and Halfter, 1993). Later, the expression of TN in specific regions of the optic nerve may restrict the distribution of myelin-forming oligodendrocytes (Bartsch *et al.*, 1994). As the axons course into the tectum, TN is highly expressed only in the target lamina (Perez and Halfter, 1993).

In cerebellar development, one of the major morphological changes is the massive migration of granule neurons over specialized glial cells (Bergmann glia) which have been shown to express TN (Chuong *et al.*, 1987; Prieto *et al.*, 1990; Bartsch *et al.*, 1992). Migration of granule cells over these glia in organotypic cultures from both the chicken and mouse is impeded by antibodies to specific regions of TN (Chuong *et al.*, 1987; Husmann *et al.*, 1992; Bartsch *et al.*, 1992). In the cerebral cortex, TN is transiently expressed at the borders of the barrel fields in layer IV of the somatosensory cortex and at the border of the patch and matrix compartments in the striatum (Crossin *et al.*, 1989; Steindler *et al.*, 1989; O'Brien *et al.*, 1992). This expression occurs during the period of dendritic sprouting, and it has been suggested that TN may restrict dendritic outgrowth to these compartments (Sheppard *et al.*, 1991; O'Brien *et al.*, 1992). In the barrel cortex, the observation that the segregation of thalamocortical axons precedes the patterned expression of TN (Jhaveri *et al.*, 1991) raises the possibility that TN expression may be affected by neural activity and is in accord with the idea that TN restricts dendritic sprouting. TN function in the nervous system is described in greater detail in Chapter 3. The effects of the different domains of TN on neurite outgrowth are described below in relation to integrin receptors and other receptors for the molecule.

CELL BIOLOGICAL STUDIES OF INTACT TN AND TN RECOMBINANT DOMAINS

A number of early studies demonstrated that TN could support the attachment of a number of primary cells and cell lines (Chiquet and Fambrough, 1984; Chiquet-Ehrismann *et al.*, 1986; Chiquet-Ehrismann *et al.*, 1988; Friedlander *et al.*, 1988; Lotz *et al.*, 1989; Prieto *et al.*, 1992). For example, in our laboratory, TN was first discovered as a molecule, called cytotactin, involved in the binding between neurons and glia (Grumet *et al.*, 1985). However, other studies were equivocal about the possibility that TN supported cell attachment. This was because TN could inhibit the spreading of both fibroblastic and neuronal cell lines plated on otherwise permissive substrates such as fibronectin (Chiquet-Ehrismann *et al.*, 1988; Friedlander *et al.*, 1988; Prieto *et al.*, 1992), a phenomenon that we termed counteradhesion (Prieto *et al.*, 1992). The predominance of adhesive or counteradhesive activities in TN appears to depend both upon the receptor repertoire on a particular cell type as well as upon the assay conditions employed.

These observations as well as the striking site-restricted distribution of TN prompted us to examine the functions of the different domains within the multidomain structure of TN (see Figure 2.1). Recombinant fragments for various domains of the protein have been prepared by ourselves (Prieto *et al.*, 1992; Prieto *et al.*, 1993; Phillips *et al.*, 1995) and others (Spring *et al.*, 1989; Aukhil *et al.*, 1993) in order to explore the effect of TN on attachment of various cell types and on neurite outgrowth. These studies have localized adhesive and counteradhesive activities of TN to specific regions of the molecule. A summary of the biological activities and binding to cell surface receptors for each of the TN domains is presented in Table 2.1.

A detailed analysis of the binding properties of each of the isolated TN domains was performed using a variety of cell types (Prieto *et al.*, 1992). In

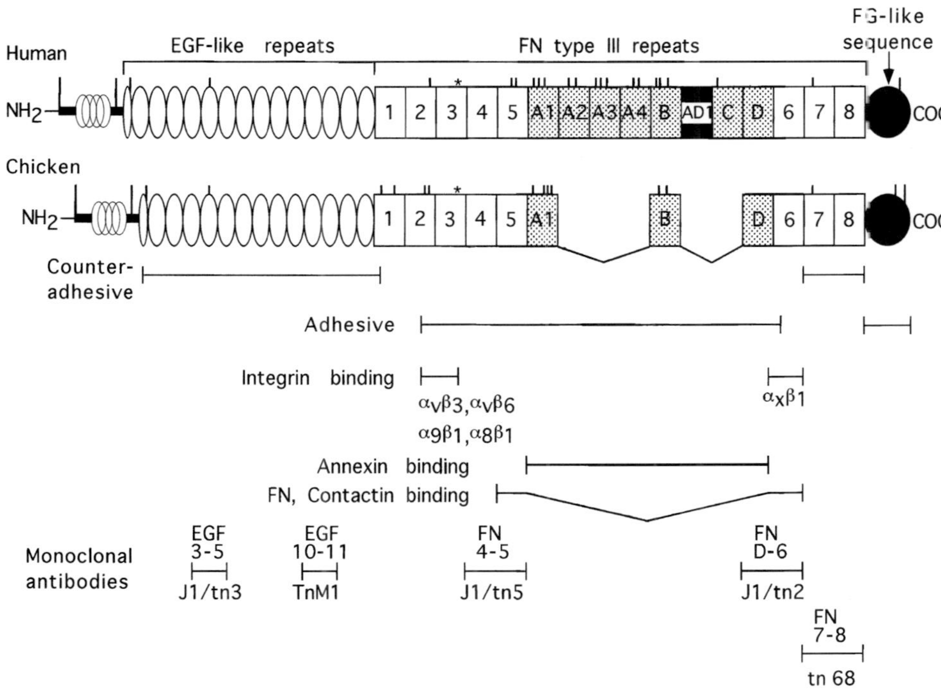

Figure 2.1 Schematic model of the domain structure of tenascin.
Schematic representations of the domains in human and chicken tenascin are shown. The shaded boxes represent fibronectin type III repeats that are alternatively spliced. Shown below are the regions where biological activities have been mapped using recombinant fragments or monoclonal antibodies.

Table 2.1 Cellular Functions and Biochemical Interactions of TN Domains

TN domain	*Function*
EGF repeats	cell rounding, counteradhesion, decreased intracellular pH
FN repeat 3	cell attachment and flattening; integrin binding ($\alpha v\beta 3$, $\alpha v\beta 6$, $\alpha 9\beta 1$, $\alpha 8\beta 1$); increased intracellular pH
FN repeats A-D	annexin II binding, enhancement of cell migration
FN repeats 5–6	fibronectin binding, contactin binding
FN repeat 6	$\beta 1$ integrin binding, promotion of central and peripheral neurite outgrowth
FN repeats 7–8	counteradhesion, cell rounding, cell attachment of trypsinized cells; decreased intracellular pH
Fibrinogen domain	cell attachment, integrin binding, proteoglycan binding, increased intracellular pH, $\beta 1$ integrin binding

Refer to Figure 2.1 for nomenclature and localization within the TN protein structure of the various domains.

contrast to early studies that identified a single cell binding site and a repulsive site (Friedlander *et al.*, 1988; Spring *et al.*, 1989), at least four non-overlapping sites on the molecule were found to interact with the cell surface (Prieto *et al.*, 1992). Two of the sites are adhesive and support strong cell attachment but show different effects on cell morphology. A region spanning the FN III repeats 2–6 promoted cell attachment and spreading and the Fg homology region promoted attachment but not spreading of fibroblasts and other cell types. Two other regions containing the EGF repeats and FN III repeats 7–8 (Figure 2.1) were found to be counteradhesive. For example, treatment of cells with either of these two fragments before plating decreased cell attachment and flattening on fibronectin (Prieto *et al.*, 1992). This finding is significant inasmuch as fibronectin and TN are coexpressed in a number of tissues where cell binding to fibronectin could be modulated (Crossin *et al.*, 1986). Cell attachment to one of the counteradhesive fragments (FN III repeats 7–8) was sensitive to the method of preparation of the single cell suspension. For this fragment, a unique TN binding site was uncovered when cells were prepared by trypsin treatment but not when cells were prepared with EDTA; counteradhesive activity was observed in cells prepared by either method. The attachment site found on trypsinized cells appears to be the one previously identified as a cell binding site (Spring *et al.*, 1989). Taken together the results suggested that multiple sites within TN mediate its adhesive and counteradhesive effects on cells. These effects are likely to be mediated by multiple, specific cell surface receptors, some of which may be differentially expressed on different cell types, as discussed further in the section below on TN receptors.

EFFECTS OF TN ON CELLULAR PHYSIOLOGY AND SIGNALLING

The dramatic effects of TN and its fragments on cell morphology suggested that this protein might have effects on cell proliferation, a process which is intimately tied to cell morphology, at least in cell culture. Early studies using 3T3 fibroblasts as well as other cells including primary glial cells and gliomas, demonstrated that these cells were inhibited in their proliferation for at least one cell cycle time in the presence of soluble TN (Crossin, 1991). In these experiments, a single cell suspension was plated onto tissue culture-treated plastic dishes in the presence or absence of soluble TN and proliferation was measured by the incorporation of tritiated thymidine. The inhibition of proliferation appeared to correlate with the ability of TN to inhibit an increase in intracellular pH (pH_i), which is known to occur upon flattening of cells on a substratum or upon growth factor stimulation (reviewed in Moolenaar, 1986; Grinstein *et al.*, 1989; Schwartz, 1992). More detailed studies on the effects of TN domains on intracellular pH will be discussed briefly below. In contrast, other studies suggested that the presence of TN could actually increase cell proliferation in certain cells. For example, transformed cells from mammary tumors both attached to TN-coated substrates and showed an increase in proliferation relative to the same cells plated on other extracellular matrix proteins (Chiquet-Ehrismann *et al.*, 1986). More recently, TN has been shown to increase the proliferation of NIH 3T3 cells (End *et al.*, 1992). Under these conditions, however, soluble TN protein was added to cells that had been growing in culture for a sufficient time to become quiescent. The mechanism mediating increased mitogenesis in this study was postulated to be different from that involving growth factors. Moreover, this stimulatory effect was mimicked by a recombinant protein fragment encompassing the third through eighth fibronectin type III repeats including the three alternatively spliced repeats. In these same studies however, TN had a growth inhibitory effect on an epidermal cell line Pam 212 (End *et al.*, 1992) but the domain responsible for this activity was not mapped. The apparently discrepant results may be reconciled by considering the cell surface receptors that are available for TN binding in the different conditions employed in these assays. The differences in these two studies may also be related to the presence of specific growth factors that were different in the two experimental paradigms or to specific cell surface receptors present on only some cell types. Whichever the case, it is clear that binding of TN to cell surfaces can lead to signalling events that result in changes in cellular physiology.

To extend our initial studies and to determine whether the ability to modulate pH_i was related to particular domains of the TN protein, we correlated the effects of recombinant fragments on pH_i with their effects on cell morphology (Krushel *et al.*, 1994). The adhesive fragments, that is, TNfn2–6 and TNfg (see Figure 2.1), when coated onto the substrate induced an increase in pH_i in both fibroblastic and glioma cell lines that was similar to the increase observed in cells plated on fibronectin substrates. These fragments nevertheless had opposite effects on cell shape. Cells on TNfn2–6 spread out over the substrate but those on TNfg remained round. The increase in pH_i by TNfn2–6 and TNfg appears to be mediated by integrins and heparan sulfate proteoglycans, respectively. This conclusion is based on the observation that the increased pH_i in cells plated on TNfn2–6 substrate could be abolished by preincubating cells with an RGD-containing peptide, thereby inhibiting integrin function. The rise in pH_i in cells plated on TNfg could be abrogated by removing cell surface heparan sulfate by preincubating the fibroblast or glioma cells with Heparinase I or III. In contrast, the counteradhesive repeats TNegf and TNfn-7–8 kept the cells at a lower pH similar to the effect of intact TN. The ability of TN and its fragments to modulate pH_i required the activity of protein kinase C, inasmuch as drugs that modulate protein kinase C activity were able to reverse their effects.

These studies are relevant to recent experiments showing that TN is subject to degradation by matrix metalloproteinases *in vitro* (Imai *et al.*, 1994; Siri *et al.*, 1995). This suggests that proteolytic fragments with different effects on cellular physiology may exist during development, in wounding, and in disease states. The ability of TN to modulate intracellular pH can affect a variety of cellular processes. pH changes of this magnitude have been shown to affect the activity of cytoplasmic enzymes (Hyrc and Rose, 1990), intracellular trafficking (Parton *et al.*, 1991; Cosson *et al.*, 1989; Samuelson *et al.*, 1988) or affect cytoskeletal integrity through effects on the pH sensitive actin depolymerizing factor (Morgan *et al.*, 1993). Intracellular pH has also been shown to alter neural physiology (Campos de Carvalho *et al.*, 1984; Konnerth *et al.*, 1987; Davies *et al.*, 1988).

RECEPTORS FOR TN

TN, like many extracellular matrix proteins, has been found to interact with a number of cell surface receptors and with other extracellular matrix molecules. Some of the interactions have been mapped to particular TN domains. Whether any of these interactions precludes or enhances other interactions is of great interest since only certain combinations of ECM proteins and cell surface receptors are present at particular embryonic locales. Because of the diversity of the molecules that bind to TN, this section is subdivided into a section on cell surface receptors and one on extracellular matrix ligands. We focus first on the cell surface receptors as prime candidates for regulating the signalling events described in the previous section.

Cell surface receptors A number of cell surface receptors for TN have now been identified using the intact TN protein, TN peptides or recombinant TN fragments. Some have been identified using antibody blockade of cell surface receptors in actual cell attachment assays and others have been postulated to be cell surface receptors because purified cell surface components bind to purified TN in biochemical assays.

Early studies suggested that integrins, a family of heterodimeric cell surface receptors that recognize a number of extracellular matrix ligands (reviewed in Hynes, 1992), were receptors for TN. This was based on the observation that chicken TN contained the RGD tripeptide sequence, which is present in a number of extracellular matrix proteins and is involved in binding to integrins (reviewed in Yamada, 1991). This was questioned, however, when it was discovered that the RGD sequence in TN was not conserved across species: it is present in chicken (Jones *et al.*, 1989) and human (Gulcher *et al.*, 1989) but is not found in mouse, newt, and pig homologues where the corresponding sequences are RVD (Weller *et al.*, 1991), RGL (Onda *et al.*, 1991), and RAD (Nishi *et al.*, 1991), respectively. Extensive studies have now documented that a number of integrins on several different cell types can function as cell surface receptors for TN and its specific domains (Bourdon and Ruoslahti, 1989; Joshi *et al.*, 1993; Sriramarao *et al.*, 1993; Prieto *et al.*, 1993; Yokosaki *et al.*, 1994; Phillips *et al.*, 1995). Our studies showed that $\alpha v\beta 3$ and, another αv integrin, presumably $\alpha v\beta 6$, could function as receptors for the third fibronectin type III repeat in both the mouse and chicken TN protein. This was shown with FC9 cells that express αv integrin subunits but not $\beta 3$ subunits, and with clones of these cells into which the $\beta 3$ integrin subunit cDNA had been transfected (Leavesley *et al.*, 1992). Binding of these cells to recombinant proteins comprising the third fibronectin type III repeat of the chicken and mouse TN proteins with and without mutations in the RGD sequence was assessed. Both the mouse and chick repeats were shown to bind to $\alpha v\beta 3$, but mutagenesis studies suggested that the requirement for RGD was not as stringent as it was for cell binding to other extracellular matrix proteins. For as example, when the chicken RGD was mutated to RAD, binding to the third fibronectin repeat in TN was not affected; mutation to RAA however severely decreased binding. Cells bound to the mouse repeat and the binding was not significantly affected when the mouse RVD sequence was mutated to RGD. In contrast, other studies have shown that mutation of the RGD sequence to RAD in the extracellular matrix protein vitronectin completely abolishes vitronectin binding to $\alpha v\beta 3$ (Cherny *et al.*, 1993). This suggests that recognition of different ligands by the same integrin can have distinct sequence requirements that may involve sequences flanking the RGD sequence.

Studies using human endothelial cells have also shown that the RGD-containing repeat of human TN can bind to $\alpha v\beta 3$ (Sriramarao *et al.*, 1993). More recent studies have demonstrated that $\alpha 9\beta 1$, an integrin that is strongly expressed on human lung cells, can attach to the third fibronectin repeat of chicken TN. TN is present in lung tissue of several species (Crossin *et al.*, 1986; Prieto *et al.*, 1990; Koch *et al.*, 1991; Tsukamoto *et al.*, 1991; Young *et al.*, 1994) indicating that this integrin interaction with TN may be significant *in vivo*. The binding to integrin $\alpha9\beta1$ is also not strongly affected by the presence of RGD since the $\alpha9\beta1$ can also bind to the RAD and RAA mutant (Yokosaki *et al.*, 1994). Whether this indicates that another sequence outside of the RGD region is involved is currently under investigation. More recent studies have shown that $\alpha8\beta1$, which is present at high levels in nervous tissue (Bossy *et al.*, 1991), binds to the third fibronectin repeat of chicken TN (Schnapp *et al.*, 1995). It is clear that at least four different integrins which are differentially distributed on various cells in development and adult animals can attach to TN through the third RGD-containing fibronectin repeat. Because all of these integrins may have additional extracellular matrix ligands, the question of which are utilized *in vivo* is of great interest.

These studies have also identified $\beta1$ integrin-dependent cell attachment to intact TN that is not mediated by the third RGD-containing repeat (Sriramarao *et al.*, 1993; Prieto *et al.*, 1993; Phillips *et al.*, 1995). One study indicated that endothelial cells bound to an unidentified region of TN via the $\alpha2\beta1$ integrin (Sriramarao *et al.*, 1993). Localization of one of these $\beta1$ integrin binding domains has been clarified by other studies on fibroblast and neuronal attachment to various domains in TN. Inasmuch as $\beta1$ integrins have been strongly implicated in neurite outgrowth (reviewed in Reichardt and Tomaselli, 1991), we tested the TN fragments for their ability to support neuron attachment and neurite outgrowth to identify the regions of TN responsible for $\beta1$ binding activity in both central and peripheral neurons. A fragment spanning the FN type III repeats 4 – 6, previously shown to sustain attachment of fibroblasts (Prieto *et al.*, 1993), was also found to support the cell attachment and outgrowth of neurites from chick dorsal root ganglion (DRG) neurons (Phillips *et al.*, 1995). To determine precisely the FN type III repeat responsible for this activity, we generated fusion proteins spanning FN III repeats 4 alone, 5 alone or 6 alone, 4 – 6, and 5 – 6. Neuron attachment and neurite outgrowth promoting activities were confined to fusion proteins containing the sixth repeat; neither activity was supported by TNfn4 or TNfn5. We concluded therefore that the sixth FN type III repeat in TN has the ability to support neuron attachment and neurite outgrowth.

Cellular attachment assays were used to determine whether the interaction of neural cells with TNfn6 was mediated by $\beta1$ integrins. DRG neurons bind to both TNfn3 and TNfn6 repeats as well as TNfg but not to the other recombinant fragments spanning the length of the TN protein. The binding of these neurons to TNfn6 and to a lesser extent to TNfg was inhibitable by antibodies to $\beta1$ integrins while attachment to TNfn3 was not affected. In contrast, CNS neurons bound only to TNfn6 and TNfg but not TNfn3. As observed for the DRG neurons, antibodies to $\beta1$ integrins could inhibit CNS neuron binding to TNfn6 and to a lesser extent to TNfg. Thus, interaction with TNfn6 at the surface of neurons is mediated by integrins of the $\beta1$ family, and this interaction appears to be the basis for the neurite outgrowth promotion activity found in this domain.

Although TNfn6 supported neurite outgrowth, its activity was never as great as that of intact TN. Quantitative experiments showed that TNfn6 increased the percentage of neurons that sprouted neurites versus control substrates, whereas TNfn3 increased the total neurite length per neurite bearing cell but had no effect on sprouting. When these two domains were combined on the same substrate, an increase in both the sprouting percentage and neurite length was observed, that was comparable to the activity of intact TN. In contrast to DRG neurons, CNS neurons, while able to grow neurites on TNfn6 substrates, did not grow on TNfn3 substrates and mixtures of TNfn3 and TNfn6 had no synergistic effect. For forebrain neurons, neurite length was essentially the same for all substrates that supported sprouting.

These observations suggest two conclusions: (1) CNS and PNS neurons have different subsets of integrins that are capable of recognizing specific domains of TN; and (2) neurite sprouting and neurite elongation mediated by TN are separable events and are mediated by distinct domains within TN by distinct receptors on the surface of neurons.

In addition to integrins, another cell surface receptor, the protein annexin II, has been shown to bind TN. This conclusion derives from Scatchard analysis and affinity isolation of receptors on human endothelial cells using recombinant fragments of human TN. These studies have revealed that the alternatively spliced repeat, containing FN repeats A through D in the human (see Figure 2.1), can bind to annexin II, a cytoskeletal protein which has also been shown to be present on the surface of a variety of cell types (Chung and Erickson, 1994). No other domains were found to attach to annexin. The finding of a receptor exclusive for the alternatively spliced domain is significant inasmuch as TN molecules containing this domain have been shown to be associated with a number of areas of cell migration during development including the cornea (Kaplony *et al.*, 1991) and the cerebellum (Chuong *et al.*, 1987; Prieto *et al.*, 1990). It may also be relevant that in endothelial cells this domain as well as intact TN itself has been shown to disperse focal adhesions, the close contacts between the cell surface and extracellular matrix of cultured cells (Murphy-Ullrich *et al.*, 1991).

Another cell surface protein, the neural cell adhesion molecule contactin/F11, has been shown in biochemical assays to bind to TN (Zisch *et al.*, 1992). Contactin is a molecule present on the surface of neurons and linked by a phosphatidylinositol anchor (Ranscht, 1988). It is therefore a good candidate for mediating interactions between neurons and glial-derived TN, although the necessary cell binding assays have not yet been performed. The interaction between TN and contactin has been shown in several types of biochemical assay to be mediated by the immunoglobulin-like domains in the contactin protein. Interaction of contactin with TN appears to be selectively mediated by the smallest unspliced isoform of TN suggesting that the binding site for this cell-surface protein is in the domain spanning the fifth and sixth fibronectin type III repeats (see Figure 2.1).

Extracellular Matrix Ligands. A number of studies have shown that fibronectin binds to TN (Chiquet *et al.*, 1991; Hoffman *et al.*, 1988) and some authors have postulated (Chiquet-Ehrismann *et al.*, 1988; Lightner and Erickson, 1990) that the binding between TN and fibronectin is responsible for the regulation of cell behavior by TN on fibronectin substrates. Although this mechanism cannot be formally ruled out, several studies provide alternative possibilities. For example, the presence of TN inhibits cell attachment and flattening on substrates with which the molecule does not interact, such as tissue culture plastic (Spring *et al.*, 1989; Crossin, 1991). Moreover, binding studies with isolated proteolytic or recombinant fragments of TN do not support the hypothesis that interaction of TN with fibronectin mediates the counteradhesive effect of TN. For example, the smallest splice variant of chicken TN has been shown to bind most strongly to intact fibronectin (Chiquet *et al.*, 1991). Inclusion of the three alternatively spliced repeats weakens this binding. This observation suggests that the site for interaction with fibronectin occurs between the fifth and sixth fibronectin type III repeat. A pronase fragment of native TN containing the sixth, seventh, and eighth fibronectin repeats and part of the fibrinogen domain (Chiquet *et al.*, 1991) and lambda fusion proteins including the EGF repeats (Spring *et al.*, 1989) both inhibited cell spreading on fibronectin substrates despite the fact that neither fragment contains the fibronectin binding region. Other studies showed that preincubation of cells with recombinant fragments of TN, including EGF repeats, and fibronectin type III 7–8, was able to inhibit attachment and spreading of cells on a fibronectin-coated substrate even after the excess proteins have been washed away (Prieto *et al.*, 1992). Together, these studies indicate that the counteradhesive effects of TN are more likely to be mediated via distinct receptors at the cell surface rather than by binding of TN to fibronectin and masking of cell binding sites on fibronectin.

Other binding studies have demonstrated that several collagens and heparin bind to TN (Faissner and Kruse, 1990; Aukhil *et al.*, 1993; Weber *et al.*, 1995). Early studies identified a soluble chondroitin sulfate proteoglycan (CTB proteoglycan), synthesized by neurons, that bound to isolated TN in biochemical assays; the binding to TN was retained even after chondroitinase treatment removed the chondroitin sulfate side chains (Hoffman and Edelman, 1987; Hoffman *et al.*, 1988). A heparin binding domain has been localized within the TN protein structure. A pronase fragment from the distal domains of intact TN including the last three FN type III repeats and most, if not all, of the fibrinogen domain bound to heparin-agarose whereas more proximal fragments did not (Chiquet *et al.*, 1991). Recombinant fragments have also been used to localize further the heparin binding domain. Heparin binding activity was mapped to two different segments, one containing the fifth fibronectin type III repeat (Weber *et al.*, 1995) and the other containing the fibrinogen domain (Aukhil *et al.*, 1993).

Proteoglycans are also found on the surface of many cell types and therefore these molecules may also function as cell surface receptors for TN. For example, fibroblast binding to the fibrinogen domain was strongly inhibited in the presence of soluble heparin or by pretreating the cells with heparatinase. As mentioned above, increased intracellular pH mediated by TNfg was abrogated by pretreatment of cells with heparitinase (Krushel *et al.*, 1994). All of these findings support the possibility that the fibrinogen domain in TN binds to heparan sulfate-containing extracellular matrix and cell surface proteoglycans. Like other matrix proteins that bind to heparin, TN has also been shown to bind to sulfatides (Crossin and Edelman, 1992). Consistent with these observations, TN has been found to interact with the cell surface proteoglycan syndecan when it is isolated from tooth mesenchyme, but only weak interaction was observed when it was isolated from mammary epithelial cells, a source of syndecan having a lower content of heparan sulfate in its side chains. TN has also been shown to bind to other cell surface chondroitin sulfate-containing proteoglycans, a notable one of which has a tyrosine phosphatase domain its intracellular domain and may therefore participate in TN-mediated cell signalling (Barnea *et al.*, 1994). These and other proteoglycan interactions are described in greater detail in Chapter 7.

THE TN NULL MUTANT MOUSE

Given the site-restricted distribution of TN and the variety of cellular processes affected by the protein itself or affected by blocking the function of TN with antibodies, it initially seemed surprising that targeted deletions of the TN gene (Saga *et al.*, 1992) resulted in mice that appeared to be grossly normal. At the time of the original report, a detailed investigation of the nervous system and behavioral analyses had not been carried out. This observation is similar to the results of an increasing number of gene deletion experiments that have revealed little or no apparent phenotypic consequences upon knock-out of a gene considered to be developmentally significant (Erickson, 1993). The general usefulness of knock-out experiments in dissecting the role of proteins in development has recently been challenged (Crossin, 1994; Routtenberg, 1995). It is possible, for example, that when one member of a gene family is deleted, other family members or other proteins may compensate for its loss. This has been documented in the case of the *myoD* deletion in which *myf*-5 expression, which is normally downregulated when *myoD* is expressed, was retained when *myoD* was deleted (Rudnicki *et al*, 1992). When both *myoD* and *myf*-5 were deleted (Rudnicki *et al.*, 1993), however, mice were born alive but immobile and died shortly after birth. The double knockout mice had no skeletal muscle or skeletal muscle marker proteins. The possibility of compensatory changes and the long term effects of the null mutation for TN in response to infection, disease, and aging remain to be explored.

Until these matters have been investigated, one cannot conclude that the gross appearance of normality in the null mutation indicates that TN has no function. Indeed recent studies have suggested that the phenotype of the TN null mutant mice is not completely normal. One such study examining the development of the barrel cortex in mutant mice has shown that the development of the barrel field is normal in TN knockout mice, but the response of glial cells after wounding in the cortex is abnormal in these animals (Steindler *et al.*, 1995). Other preliminary studies in the TN knockout mice have suggested that their motor coordination is abnormal and that gut innervation may be abnormal. These observations would be consistent with a role for TN in affecting physiological processes.

EFFECTS OF TN BINDING ON GENE TRANSCRIPTION

Given that TN is an adhesive molecule with a highly place-dependent expression pattern, it is important to examine the changes in gene expression that result from its binding to the cell surface. A number of studies have indicated that changes in gene expression occur in response both to cell-cell and to cell-substrate adhesion. Various integrins have recently been recognized as signal transducing receptors (Schwartz, 1992), and there is growing evidence that integrins, through their effects on cell shape and signal transduction, may induce changes in gene expression (Adams and Watt, 1993). Cell attachment through β1 integrins, together with hormonal stimulation, is required for β-casein synthesis by mammary epithelial cells (Streuli *et al.*, 1991); the region of the β-casein gene that mediates this effect has been identified (Schmidhauser *et al.*, 1992). Attachment through α5 integrins has been shown to decrease the synthesis of matrix-degrading metalloproteinases (Damsky and Werb, 1992; Werb *et al.*, 1989). Several other studies have implicated integrins as mediators of altered gene expression upon binding of cells to a variety of ECM substrates (Ben-Ze'ev, 1986; DiPersio *et al.*, 1991; Haskill *et al.*, 1991; Liu *et al.*, 1991; Tremble *et al.*, 1994). For example, compared with fibroblasts plated on fibronectin alone, fibroblastic cells plated on a mixture of fibronectin and TN upregulated the synthesis of four gene products: collagenase, stromelysin, a 92 kDa gelatinase, and c-fos (Tremble *et al.*, 1994). This effect on matrix metalloproteinase gene expression was reversed in the presence of monoclonal antibody TN 68 that recognizes an epitope within the seventh and eighth fibronectin type III repeat in TN. No effect was seen with another monoclonal antibody TN M1 that recognizes an epitope in the EGF domains of TN. These studies suggested that the cell shape changes mediated by the presence of TN could lead to changes in gene expression in the target cells.

We further explored the question of whether the binding of TN to chick brain neurons altered the expression of other genes by using a subtractive hybridization strategy similar to one used previously (Mauro *et al.*, 1994) to reveal changes in gene expression brought about by N-CAM-mediated aggregation of chick embryo brain (CEB) cells. The subtractive hybridization cloning strategy involves removing sequences common to two cDNA libraries, made from cells incubated with or without TN, thereby enriching for sequences whose expression is either increased or decreased as a result of binding of TN to the cell surface. Poly A$^+$ RNA was prepared from TN-treated CEB cells and from untreated cells. Directional cDNA libraries were then prepared and subtractive hybridization was carried out as previously described (Mauro *et al.*, 1994).

The subtracted cDNA libraries were analyzed by hybridization and sequence analysis. Duplicate filters made from these libraries were initially screened using radiolabelled subtracted cDNA as a probe to identify clones that were enriched in one library relative to the other. They were then subsequently screened with radiolabelled first strand cDNA probes derived from chick brain cells that were either treated with TN or were untreated, as in the initial experiments. cDNA clones in the subtracted libraries that had a different pattern of hybridization with these two first strand cDNA probes were further processed (Mauro *et al.*, 1994) and sequenced.

Three hundred cDNA clones were sequenced and compared to known sequences using the FASTA program (Genetics Computer Group, Version 7.2-UNIX) to scan the GenBank (release 80), EMBL (release 35), and Swiss-Prot (release 26) databases. Of these clones, approximately 6% encoded known mRNAs (Table 2.2). This reflects both the complexity of the subtracted cDNA libraries and presumably the large number of mRNAs in the brain that have not yet been characterized.

A number of cDNA clones encoding CAMs, ECM molecules, and ribosomal proteins were enriched in the library made from TN-treated cells (Table 2.2). One of the cDNA clones enriched in the TN-treated library encodes neurofascin, a cell adhesion molecule that is a member of the immunoglobulin superfamily (Rathjen *et al.*, 1987). The cDNA for the chicken homologue of the opioid-binding CAM, OB-CAM, was also found to be enriched in this library. Another cDNA clone encoding claustrin, an extracellular keratan sulfate proteoglycan with counteradhesive properties similar to those of TN, was found to be enriched in the TN-untreated library and was therefore apparently down-regulated by TN treatment. Using this approach, we also identified what appears to be a new molecule that is very similar in sequence to claustrin. The finding of claustrin in the untreated library is particularly interesting because of the functional similarities between claustrin and TN: both are ECM molecules with counteradhesive properties (McCabe and Cole, 1992). Claustrin is a 320 kDa keratan sulfate proteoglycan that has been demonstrated to inhibit both neurite outgrowth and cell attachment on laminin and N-CAM substrates (Cole and McCabe, 1991). Structurally, claustrin is related to the mouse microtubule-associated protein MAP1B. The observations indicating differential expression of claustrin and a claustrin-related sequence, of neurofascin and of OB-CAM, support the possibility that the expression of genes for various adhesion molecules may be altered by binding of TN to the cell surface.

Table 2.2 Summary of Changes in Gene Expression in Cells Interacting with TN.

	mRNAs enriched in CT-Treated Brain Cells	**mRNAs enriched in Untreated Brain Cells**
CAMs/SAMs	neurofascin OB-CAM	claustrin novel-claustrin-like protein
Gene control	novel RING-zinc finger protein BROMO-domain protein translationally controlled tumor protein	
Ribosomal proteins	L3, L19, L27, L29, L38, S2	S1
Miscellaneous	Fatty acid binding protein tubulin	glutathione-S-transferase MARKS-kinase substrate

Primary neurons isolated from embryonic day 8 chick embryo brains prepared according to standard protocols (Brackenbury *et al.*, 1984) were plated onto polylysine-coated tissue culture dishes at a density of 5 X 10^7 cells per dish for 20 minutes to allow the cells to attach. The medium was changed and the cells were then incubated for 1 hour in the presence or absence of 15 mg/ml chicken TN (an amount previously shown to be maximal for several biological effects (Crossin, 1991; Prieto *et al.*, 1992)). In order to minimize the contribution from TN-induced morphological changes, the above conditions were established so that cell shape was identical in both treated and untreated cells (that is, it was confirmed that, throughout the incubation period, cells remained round in the presence or absence of TN).

PERSPECTIVES

It is now clear that different regions within the TN protein sequence can independently mediate specific biological responses. This occurs presumably by binding to specific cell surface receptors for each of the functional domains. The most dramatic effects, including cell rounding, lowered intracellular pH, and certain changes in gene expression, are mediated by the counteradhesive domains, the EGF repeats and FN type III repeats 7–8. Receptors specific for these counteradhesive domains have not yet been identified and they are therefore of great interest. One adhesive domain, the RGD-containing third FN type III repeat, binds to multiple integrin receptors which may be simultaneously present on the cell surface. This is further complicated by the fact that other ECM molecules can be present in the same sites as TN and may bind to the same receptors. The differential affinity of such molecular interactions may therefore determine which receptor is occupied and therefore how a particular cell responds in the presence of TN. Binding to TN may be similarly modulated by the presence or absence of the alternatively spliced FN type III domains. For example, annexin II binds to the alternatively spliced domains A–D. TN isoforms lacking these domains bind more avidly to contactin and to fibronectin. These findings suggest that alternative splicing of TN may select for different receptor repertoires.

Whether binding of a particular ECM protein or surface receptor to TN affects the subsequent binding to other molecules deserves further investigation. For example, intact TN has counteradhesive properties similar to the isolated EGF or FN7–8 repeats. Why are the adhesive sites within the protein not available or functional in the context of the intact molecule? It has been suggested, for example, based upon crystallographic studies (Leahy *et al.*, 1992), that the RGD sequence could be hidden in intact TN. The ability of TN to mediate its adhesive and counteradhesive effects may itself be modulated by its interaction with other extracellular matrix proteins or with specific cell surface receptors. Clearly, the expression of domain specific receptors is one other parameter that can alter the responses of cells to the presence of TN or its fragments. The activity of intact TN would therefore represent the summation of its adhesive and counteradhesive activities in the context of the repertoire of cell surface receptors on a particular cell type. Further analysis of the cell interactions that mediate the adhesive and counteradhesive effects of TN should lead to a better understanding of the regulation of a variety of cellular processes important to development and to disease, including cell proliferation, migration, neurite sprouting and outgrowth, cell differentiation, and differential gene expression.

REFERENCES

1. Adams, J.C., and Watt, F.M. (1993) Regulation of development and differentiation by the extracellular matrix. *Development*, **117**, 1183–1198.
2. Aufderheide, E., and Ekblom, P. (1988) Tenascin during gut development: appearance in the mesenchyme, shift in molecular forms, and dependence on epithelial-mesenchymal interactions. *J. Cell Biol.*, **105**, 2341–2349.
3. Aukhil, I., Joshi, P., Yan, Y., and Erickson, H.P. (1993) Cell- and heparin-binding domains of the hexabrachion arm identified by tenascin expression proteins. *J. Biol. Chem.*, **268**, 2542–2553.
4. Barnea, G., Grumet, M., Milev, P., Silvennoinen, O., Levy, J.B., Sap, J., and Schlessinger, J. (1994) Receptor tyrosine phosphatase beta is expressed in the form of proteoglycan and binds to the extracellular matrix protein tenascin. *J. Biol. Chem.*, **269**, 14349–14352.
5. Bartsch, S., Bartsch, U., Dories, U., Faissner, A., Weller, A., Ekblom, P., and Schachner, M. (1992) Expression of tenascin in the developing cerebellar cortex. *J. Neurosci.*, **12**, 736–749.
6. Bartsch, U., Faissner, A., Trotter, J., Dorries, U., Bartsch, S., Mohajeri, H., and Schachner, M. (1994) Tenascin demarcates the boundary between the myelinated and nonmyelinated part of retinal ganglion cell axons in the developing and adult mouse. *J. Neurosci.*, **14**, 4756–4768.

7. Ben-Ze'ev, A. (1986) The relationship between cytoplasmic organization, gene expression and morphogenesis *TIBS*, **11**, 478–481.
8. Bossy, B., Bossy-Wetzel, E., and Reichardt, L.F. (1991) Characterization of the ntegrin alpha 8 subunit: a new integrin beta 1-associated subunit, which is prominently expressed on axons and on cells in contact with basal laminae in chick embryos. *EMBO J.*, **10**, 2375–2385.
9. Bourdon, M.A., and Ruoslahti, E. (1989) Tenascin mediates cell attachment through an RGD-dependent receptor. *J. Cell Biol.*, **108**, 1149–1155.
10. Brackenbury, R., Greenberg, M.E., and Edelman, G.M. (1984) Phenotypic changes and loss of N-CAM mediated adhesion in transformed embryonic chicken retinal cells. *J. Cell Biol.*, **99**, 1944–1954.
11. Bristow, J., Tee, M.K., Gitelman, S.E., Mellon, S.H., and Miller, W.L. (1993) Tenascin-X: A novel extracellular matrix protein encoded by the human XB gene overlapping P450c21B. *J. Cell Biol.*, **122**, 265–278.
12. Bronner-Fraser, M. (1988) Distribution and function of tenascin during cranial neural crest development in the chick. *J. Neurosci. Res.*, **21**, 135–147.
13. Campos de Carvalho, A., Spray, D.C., and Bennett, M.V.L. (1984) pH dependence of transmission at electrotonic synapses of the crayfish septate axon. *Brain Res.*, **321**, 279–286.
14. Cherny, R.C., Honan, M.A., and Thiagarajan, P. (1993) Site-directed mutagenesis of the arginine-glycine-aspartic acid in vitronectin abolishes cell adhesion. *J. Biol. Chem.*, **268**, 9725–9729.
15. Chiquet, M., and Fambrough, D.M. (1984) Chick myotendinous antigen. I. A monoclonal antibody as a marker for tendon and muscle morphogenesis. *J. Cell Biol.*, **98**, 1926–1936.
16. Chiquet, M., Wehrle-Haller, B., and Koch, M. (1991) Tenascin (cytotactin): an extracellular matrix protein involved in morphogenesis of the nervous system. *Neuroscience*, **3**, 341–350.
17. Chiquet-Ehrismann, R., Kalla, P., Pearson, C.A., Beck, K., and Chiquet, M. (1988) Tenascin interferes with fibronectin action. *Cell*, **53**, 383–390.
18. Chiquet-Ehrismann, R., Mackie, E.J., Pearson, C.A., and Sakakura, T. (1986) Tenascin: an extracellular matrix protein involved in tissue interactions during fetal development and oncogenesis. *Cell*, **47**, 131–139.
19. Chung, C.Y., and Erickson, H.P. (1994) Cell surface annexin II is a high affinity receptor for the alternatively spliced segment of Tenascin-C. *J. Cell Biol.*, **126**, 539–548.
20. Chuong, C.-M., Crossin, K.L., and Edelman, G.M. (1987) Sequential expression and differential function of multiple adhesion molecules during the formation of cerebellar cortical layers. *J. Cell Biol.*, **104**, 331–342.
21. Cole, G.J., and McCabe, C.F. (1991) Identification of a developmentally regulated keratan sulfate proteoglycan that inhibits cell adhesion and neurite outgrowth. *Neuron*, **7**, 1007–1018.
22. Cosson, P., deCurtis, I., Pouyssegur, J., Griffiths, G., and Davoust, J. (1989) Low cytoplasmic pH inhibits endocytosis and transport from the trans-Golgi network to the cell surface. *J. Cell Biol.*, **108**, 377–387.
23. Crossin, K.L. (1991) Cytotactin binding: Inhibition of stimulated proliferation and intracellular alkalinization in fibroblasts. *Proc. Natl. Acad. Sci. USA*, **88**, 11403–11407.

24. Crossin, K.L. (1994) The functional role of cytotactin/tenascin in morphogenesis: a modest proposal. *Persp. Dev. Neurobiol.*, **2**, 21–32.
25. Crossin, K.L., and Edelman, G.M. (1992) Specific binding of cytotactin to sulfated glycolipids. *J. Neurosci. Res.*, **33**, 631–638.
26. Crossin, K.L., Hoffman, S., Grumet, M., Thiery, J.-P., and Edelman, G.M. (1986) Site-restricted expression of cytotactin during development of the chicken embryo. *J. Cell Biol.*, **102**, 1917–1930.
27. Crossin, K.L., Hoffman, S., Tan, S.-S., and Edelman, G.M. (1989) Cytotactin and its proteoglycan ligand mark structural and functional boundaries in somatosensory cortex of the early postnatal mouse. *Dev. Biol.*, **136**, 381–392.
28. Damsky, C.H., and Werb, Z. (1992) Signal transduction by integrin receptors for extracellular matrix: cooperative processing of extracellular information. *Curr. Op. Cell Biol.*, **4**, 772–781.
29. Davies, N.W., Lux, H.D., and Morad, M. (1988) Site and mechanism of activation of proton-induced sodium current in chick dorsal root ganglion neurones. *J. Physiol.*, **400**, 159–187.
30. DiPersio, C.M., Jackson, D.A., and Zaret, K.S. (1991) The extracellular matrix coordinately modulates liver transcription factors and hepatocyte morphology. *Mol. Cell Biol.*, **111**, 4405–4414.
31. Duband, J.-L., Dufour, S., Hatta, K., Takeichi, M., Edelman, G.M., and Thiery, J.-P. (1987) Adhesion molecules during somitogenesis in the avian embryo. *J. Cell Biol.*, **104**, 1361–1374.
32. End, P., Panayotou, G., Entwistle, A., Waterfield, M.D., and Chiquet, M. (1992) Tenascin: a modulator of cell growth. *Eur. J. Biochem.*, **209**, 1041–1051.
33. Epperlein, H.H., Halfter, W., and Tucker, R.P. (1988) The distribution of fibronectin and tenascin along migratory pathways of the neural crest in the trunk of amphibian embryos. *Development*, **103**, 743–756.
34. Erickson, H.P. (1993) Gene knockouts of c-*src*, transforming growth factor β1, and tenascin suggest superfluous, nonfunctional expression of proteins. *J. Cell Biol.*, **120**, 1079–1081.
35. Faissner, A., and Kruse, J. (1990) J1/tenascin is a repulsive substrate for central nervous system neurons. *Neuron*, **5**, 627–637.
36. Friedlander, D.R., Hoffman, S., and Edelman, G.M. (1988) Functional mapping of cytotactin: proteolytic fragments active in cell-substrate adhesion. *J. Cell Biol.*, **107**, 2329–2340.
37. Fuss, B., Wintergerst, E.-S., Bartsch, U., and Schachner, M. (1993) Molecular characterization and in situ mRNA localization of the neural recognition molecule J1-160/180: a modular structure similar to tenascin. *J. Cell Biol.*, **120**, 1237–1249.
38. Grinstein, S., Rotin, D., and Mason, M.J. (1989) Na^+/H^+ exchange and growth factor-induced cytosolic pH changes. Role in cellular proliferation. *Biochim. Biophys. Acta*, **988**, 73–97.
39. Grumet, M., Hoffman, S., Crossin, K.L., and Edelman, G.M. (1985) Cytotactin, an extracellular matrix protein of neural and non-neural tissues that mediates glia-neuron interaction. *Proc. Natl. Acad. Sci. USA*, **82**, 8075–8079.
40. Gulcher, J.R., Nies, D.E., Marton, L.S., and Stefansson, K. (1989) An alternatively spliced region of the human hexabrachion contains a novel repeat of potential N-glycosylation sites. *Proc. Natl. Acad. Sci. USA*, **86**, 1588–1592.

41. Halfter, W., Chiquet-Ehrismann, R., and Tucker, R.P. (1989) The effect of tenascin and embryonic basal lamina on the behavior and morphology of neural crest cells *in vitro*. *Dev. Biol.*, **132**, 14–25.
42. Haskill, S., Beg, A.A., Tompkins, S.M., Morris, J.S., Yurochko, A.D., Sampson-Johannes, A., Mondal, K., Ralph, P., and Baldwin, A.S. (1991) Characterization of an immediate-early gene induced in adherent monocytes that encodes I κB-like activity. *Cell*, **65**, 1281–1289.
43. Hoffman, S., Crossin, K.L., and Edelman, G.M. (1988) Molecular forms, binding functions, and developmental expression patterns of cytotactin and cytotactin-binding proteoglycan, an interactive pair of extracellular matrix molecules. *J. Cell Biol.*, **106**, 519–532.
44. Hoffman, S., and Edelman, G.M. (1987) A proteoglycan with HNK-1 antigenic determinants is a neuron-associated ligand for cytotactin. *Proc. Natl. Acad. Sci. USA*, **84**, 2523–2527.
45. Husmann, K., Faissner, A., and Schachner, M. (1992) Tenascin promotes cerebellar granule cell migration and neurite outgrowth by different domains in the fibronectin type III repeats. *J. Cell Biol.*, **116**, 1475–1486.
46. Hynes, R.O. (1992) Integrins: versatility, modulation, and signaling in cell adhesion. *Cell*, **69**, 11–25.
47. Hyrc, K., and Rose, B. (1990) The action of v-*src* on gap junctional permeability is modulated by pH. *J. Cell Biol.*, **110**, 1217–1226.
48. Imai, K., Kusakabe, M., Sakakura, T., Nakanishi, I., and Okada, Y. (1994) Susceptibility of tenascin to degradation by matrix metalloproteinases and serine proteinases. *FEBS Letters*, **352**, 216–218.
49. Jhaveri, S., Erzurumlu, R.S., and Crossin, K.L. (1991) Barrel construction in rodent neocortex: Role of thalamic afferents versus extracellular matrix molecules. *Proc. Natl. Acad. USA*, **88**, 4489–4493.
50. Jones, F.S., Crossin, K.L., Cunningham, B.A., and Edelman, G.M. (1990) Identification and characterization of the promoter for the cytotactin gene. *Proc. Natl. Acad. Sci. USA*, **87**, 6497–6501.
51. Jones, F.S., Hoffman, S., Cunningham, B.A., and Edelman, G.M. (1989) A detailed structural model of cytotactin. Protein homologies, alternative RNA splicing, and binding regions. *Proc. Natl. Acad. Sci. USA*, **86**, 1905–1909.
52. Joshi, P., Chung, C.-Y., Aukhil, I., and Erickson, H.P. (1993) Endothelial cells adhere to the RGD domain and fibrinogen-like terminal knob of tenascin. *J. Cell Sci.*, **106**, 389–400.
53. Kaplony, A., Zimmermann, D.R., Fischer, R.W., Imhof, B.A., Odermatt, B.F., Winterhalter, K.H., and Vaughan, L. (1991) Tenascin Mr 220,000 isoform expression correlates with corneal cell migration. *Development*, **112**, 605–614.
54. Koch, M., Wehrle-Haller, B., Baumgartner, S., Spring, J., Brubacher, D., and Chiquet, M. (1991) Epithelial synthesis of tenascin at the tips of growing bronchi and graded accumulation in basement membrane and mesenchyme. *Exp. Cell Res.*, **194**, 297–300.
55. Konnerth, A., Lux, H.D., and Morad, M. (1987) Proton-induced transformation of calcium channel in chick dorsal root ganglion cells. *J. Physiol.*, **386**, 603–633.
56. Krushel, L.A., Prieto, A.L., Edelman, G.M., and Crossin, K.L. (1994) Differential effects of cytotactin/tenascin fusion proteins on intracellular pH and cell morphology. *J. Cell. Physiol.*, **161**, 508–518.

57. Leahy, D.J., Hendrickson, W.A., Aukhil, I., and Erickson, H.P. (1992) Structure of a fibronectin type III domain from tenascin phased by MAD analysis of the selenomethionyl protein. *Science*, **258**, 987–991.
58. Leavesley, D.I., Ferguson, G.D., Wayner, E.A., and Cheresh, D.A. (1992) Requirement of the integrin $\_3$ subunit for carcinoma cell spreading or migration on vitronectin and fibrinogen. *J. Cell Biol.*, **117**, 1101–1107.
59. Lightner, V.A., and Erickson, H.P. (1990) Binding of hexabrachion (tenascin) to the extracellular matrix and substratum and its effect on cell adhesion. *J. Cell Sci.*, **95**, 263–277.
60. Lightner, V.A., Gumkowski, F., Bigner, D.D., and Erickson, H.P. (1989) Tenascin/hexabrachion in human skin: biochemical identification and localization by light and electron microscopy. *J. Cell Biol.*, **108**, 2483–2493.
61. Liu, J.-K., DiPersio, C.M., and Zaret, K.S. (1991) Extracellular signals that regulate liver transcription factors during hepatic differentiation in vitro. *Mol. Cell. Biol.*, **11**, 773–784.
62. Lotz, M.M., Burdsal, C.A., Erickson, H.P., and McClay, D.R. (1989) Cell adhesion to fibronectin and tenascin: quantitative measurements of initial binding and subsequent strengthening response. *J. Cell Biol.*, **109**, 1795–1805.
63. Mackie, E.J., Halfter, W., and Liverani, D. (1988b) Induction of tenascin in healing wounds. *J. Cell Biol.*, **107**, 2757–2767.
64. Mackie, E.J., Tucker, R.P., Halfter, W., Chiquet-Ehrismann, R., and Epperlein, H.H. (1988a) The distribution of tenascin coincides with pathways of neural crest cell migration. *Development*, **102**, 237–250.
65. Matsumoto, K., Ishihara, N., Ando, A., Inoki, H., and Ikemura, T. (1992) Extracellular matrix protein tenascin-like gene found in human MHC class III region. *Immunogenetics*, **36**, 400–403.
66. Mauro, V.P., Wood, I.C., Krushel, L., Crossin, K.L., and Edelman, G.M. (1994) Cell adhesion alters gene transcription in chicken embryo brain cells and mouse embryonal carcinoma cells. *Proc. Natl. Acad. Sci. USA*, **91**, 2868–2872.
67. McCabe, C.F., and Cole, G.J. (1992) Expression of the barrier-associated proteins EAP-300 and claustrin in the developing central nervous system. *Dev. Brain Res.*, **70**, 9–24.
68. Moolenaar, W.H. (1986) Effects of growth factors on intracellular pH regulation. *Annu. Rev. Physiol.*, **48**, 363–376.
69. Morgan, T.E., Lockerbie, R.O., Minamide, L.S., Browning, M.D., and Bamburg, J.R. (1993) Isolation and characterization of a regulated form of actin depolymerizing factor. *J. Cell Biol.*, **122**, 623–633.
70. Murphy-Ullrich, J.E., Lightner, V.A., Aukhil, I., Yan, Y.Z., and Erickson, H.P. (1991) Focal adhesion integrity is downregulated by the alternatively spliced domain of human tenascin. *J. Cell Biol.*, **115**, 1127–1136.
71. Nishi, T., Weinstein, J., Gillespie, W.M., and Paulson, J.C. (1991) Complete primary structure of porcine tenascin. Detection of tenascin transcripts in adult submaxillary glands. *Eur. J. Biochem.*, **202**, 643–648.
72. O'Brien, T.F., Faissner, A., Schachner, M., and Steindler, D.A. (1992) Afferent-boundary interactions in the developing neostriatal mosaic. *Brain Res.*, **65**, 259–267.
73. Onda, H., Poulin, M.L., Tassava, R.A., and Chiu, I. (1991) Characterization of a newt tenascin cDNA and localization of tenascin mRNA during newt limb regeneration by *in situ* hybridization. *Dev. Biol.*, **148**, 219–232.

74. Parton, R.G., Dotti, C.D., Bacallao, R., Kurtz, I., Simons, K., and Prydz, K. (1991) pH-induced microtubule-dependent redistribution of late endosomes in neuronal and epithelial cells. *J. Cell Biol*, **113**, 261–274.
75. Perez, R.G., and Halfter, W. (1993) Tenascin in the developing chick visual system: Distribution and potential role as a modulator of retinal axon growth. *Dev. Biol.*, **156**, 278–292.
76. Phillips, G.R., Edelman, G.M., and Crossin, K.L. (1995) Separate cell binding sites within cytotactin/tenascin differentially promote neurite outgrowth. *Cell Adh. Comm.* **3**, 257–271.
77. Prieto, A.L., Andersson-Fisone, C., and Crossin, K.L. (1992) Characterization of multiple adhesive and counteradhesive domains in the extracellular matrix protein cytotactin. *J. Cell. Biol.*, **119**, 663–678.
78. Prieto, A.L., Edelman, G.M., and Crossin, K.L. (1993) Multiple integrins mediate cell attachment to cytotactin/tenascin. *Proc. Natl. Acad. Sci. USA*, **90**, 10154–10158.
79. Prieto, A.L., Jones, F.S., Cunningham, B.A., Crossin, K.L., and Edelman, G.M. (1990) Localization during development of alternatively-spliced forms of cytotactin by *in situ* hybridization. *J. Cell Biol.*, **111**, 685–698.
80. Ranscht, B. (1988) Sequence of contactin, a 130-kD glycoprotein concentrated in areas of interneuronal contact, defines a new member of the immunoglobulin supergene family in the nervous system. *J. Cell Biol.*, **107**, 1561–1573.
81. Rathjen, F.G., Wolff, J.M., and Chiquet-Ehrismann, R. (1991) Restrictin: a chick neural extracellular matrix protein involved in cell attachment co-purifies with the cell recognition molecule F11. *Development*, **113**, 151–164.
82. Rathjen, F.G., Wolff, J.M., Frank, R., Bonhoeffer, F., and Rutishauser, U. (1987) Membrane glycoproteins involved in neurite fasciculation. *J. Cell Biol.*, **104**, 343–353.
83. Reichardt, L.F., and Tomaselli, K.J. (1991) Extracellular matrix molecules and their receptors: Functions in neural development. *Annu. Rev. Neurosci.*, **14**, 531–570.
84. Routtenberg, A. (1995) Knockout mouse fault lines. *Nature*, **374**, 314–315.
85. Rovasio, R.A., Delouv_e, A., Yamada, K.M., Timpl, R., and Thiery, J.-P. (1983) Neural crest cell migration- Requirements for exogenous fibronectin and high cell density. *J. Cell Biol.*, **96**, 462–473.
86. Rudnicki, M.A., Braun, T., Hinuma, S., and Jaenisch, R. (1992) Inactivation of MyoD in mice leads to up-regulation of the myogenic HLH gene Myf-5 and results in apparently normal muscle development. *Cell*, **71**, 383–390.
87. Rudnicki, M.A., Schnegelsberg, P.N., Stead, R.H., Braun, T., Arnold, H.H., and Jaenisch, R. (1993) MyoD or Myf-5 is required for the formation of skeletal muscle. *Cell*, **75**, 1351–1359.
88. Saga, Y., Yagi, T., Ikawa, Y., Sakakura, T., and Aizawa, S. (1992) Mice develop normally without tenascin. *Genes & Dev.*, **6**, 1821–1831.
89. Samuelson, A.C., Stockert, R.J., Novikoff, A.B., Novikoff, P.M., Saez, J.C., Spray, D.C., and Wolkoff, A.W. (1988) Influence of cytosolic pH on receptor-mediated endocytosis of asialoorosomucoid. *Am. J. Physiol.*, **23**, C829–C838.
90. Schmidhauser, C., Casperson, G.F., Myers, C.A., Sanzo, K.T., Bolten, S., and Bissell, M.J. (1992) A novel transcriptional enhancer is involved in the prolactin- and extracellular matrix-dependent regulation of β-casein gene expression. *Mol. Biol. Cell*, **3**, 699–709.

91. Schnapp, L.M., Hatch, N., Ramos, D.M., Klimanskaya, I.V., Sheppard, D., and Pytela, R. (1995) The human integrin $\alpha 8\beta 1$ functions as a receptor for tenascin, fibronectin, and vitronectin, *J. Biol. Chem.*, **270**, in press.
92. Schwartz, M.A. (1992) Transmembrane signalling by integrins. *Trends Cell Biol.*, **2**, 304–308.
93. Sheppard, A.M., Hamilton, S.K., and Pearlman, A.L. (1991) Changes in the distribution of extracellular matrix components accompany early morphogenetic events of mammalian cortical development. *J. Neurosci.*, **11**, 3928–3942.
94. Siri, A., Knäuper, V., Veirana, N., Caocci, F., Murphy, G., and Zardi, L. (1995) Different susceptibility of small and large human tenascin-C isoforms to degradation by matrix-metalloproteinases. *J. Biochem.*, (in press).
95. Spring, J., Beck, K., and Chiquet-Ehrismann, R. (1989) Two contrary functions of tenascin: dissection of the active sites by recombinant tenascin fragments. *Cell*, **59**, 325–334.
96. Sriramarao, P., Mendler, M., and Bourdon, M.A. (1993) Endothelial cell attachment and spreading on human tenascin is mediated by $\alpha 2\beta 1$ and $\alpha v\beta 3$ integrins. *J. Cell Sci.*, **105**, 1001–1012.
97. Steindler, D.A., Cooper, N.G.F., Faissner, A., and Schachner, M. (1989) Boundaries defined by adhesion molecules during development of the cerebral cortex: The J1/tenascin glycoprotein in the mouse somatosensory cortical barrel field. *Dev. Biol.*, **131**, 243–260.
98. Steindler, D.A., Settles, D., Erickson, H.P., Laywell, E.D., Yoshiki, A., Faissner, A., and Kusakabe, M. (1995) Tenascin knockout mice: Barrels, boundary molecules, and glial scars. *J. Neurosci.*, **15**, 1971–1983.
99. Stern, C.D., Norris, W.E., Bronner-Fraser, M., Carlson, G.J., Faissner, A., Keynes, R.J., and Schachner, M. (1989) J1/tenascin-related molecules are not responsible for the segmented pattern of neural crest cells or motor axons in the chick embryo. *Development*, **107**, 309–319.
100. Streuli, C.H., Bailey, N., and Bissell, M.J. (1991) Control of mammary epithelial differentiation: basement membrane induces tissue-specific gene expression in the absence of cell-cell interaction and morphological polarity. *J. Cell Biol.*, **115**, 1385–1395.
101. Tan, S.-S., Crossin, K.L., Hoffman, S., and Edelman, G.M. (1987) Asymmetric expression in somites of cytotactin and its proteoglycan ligand is correlated with neural crest cell distribution. *Proc. Natl. Acad. Sci. USA*, **84**, 7977–7981.
102. Tan, S.S., Prieto, A.L., Newgreen, D.F., Crossin, K.L., and Edelman, G.M. (1991) Cytotactin expression in somites after dorsal neural tube and neural crest ablation in chicken embryos. *Proc. Natl. Acad. Sci. USA*, **88**, 6398–6402.
103. Tremble, P., Chiquet-Ehrismann, R., and Werb, Z. (1994) The extracellular matrix ligands fibronectin and tenascin collaborate in regulating collagenase gene expression in fibroblasts. *Mol. Biol. Cell*, **5**, 439–453.
104. Tsukamoto, T., Kusakabe, M., and Saga, Y. (1991) In situ hybridization with non-radioactive digoxigenin-11-UTP-labeled cRNA probes: localization of developmentally regulated mouse tenascin mRNAs. *Int. J. Devel. Biol.*, **35**, 25–32.
105. Tucker, R., and McKay, S.E. (1991) The expression of tenascin by neural crest cells and glia. *Development*, **112**, 1031–1039.

106. Weber, P., Zimmermann, D.R., Winterhalter, K.H., and Vaughan, L. (1995) Tenascin-C binds heparin by its fibronectin type III domain five. *J. Biol. Chem.*, **270**, 4619–4623.
107. Weller, A., Beck, S., and Ekblom, P. (1991) Amino acid sequence of mouse tenascin and differential expression of two tenascin isoforms during embryogenesis. *J. Cell Biol.*, **112**, 355–362.
108. Werb, Z., Tremble, P.M., Behrendtsen, O., Crowley, E., and Damsky, C.H. (1989) Signal transduction through the fibronectin receptor induces collagenase and stromelysin gene expression. *J. Cell Biol.*, **109**, 877–889.
109. Yamada, K.M. (1991) Adhesive recognition sequences. *J. Biol. Chem.*, **266**, 12809–12812.
110. Yokosaki, Y., Palmer, E.L., Prieto, A.L., Crossin, K.L., Bourdon, M.A., Pytela, R., and Sheppard, D. (1994) The integrin $\alpha 9\beta 1$ mediates cell attachment to a non-RGD site in the third fibronectin type III repeat of tenascin. *J. Biol. Chem.*, **269**, 26691–26696.
111. Young, S.L., Chang, L.Y., and Erickson, H.P. (1994) Tenascin-C in rat lung: distribution, ontogeny and role in branching morphogenesis. *Devel. Biol.*, **161**, 615–625.
112. Zisch, A.H., D'Alessandri, L., Ranscht, B., Falchetto, R., Winterhalter, K.H., and Vaughan, L. (1992) Neuronal cell adhesion molecule contactin/F11 binds to tenascin via its immunoglobulin-like domains. *J. Cell Biol.*, **119**, 203–213.

3. THE TENASCIN GENE FAMILY IN DEVELOPMENT AND PLASTICITY OF NEURAL TISSUES

ANDREAS FAISSNER

Department of Neurobiology, University of Heidelberg, INF 364, D-69120 Heidelberg, Germany (FRG)

It is widely accepted that glial cells play an important role in the regulation of neural pattern formation, for example by guiding migrating neuroblasts and growth cones to their final destinations. The last years have shown that, in addition to providing a supportive environment in many cases, astro- and also oligodendroglia may impose restrictions and constraints on the pathways of migrating neurons and/or of extending axonal projections. It has become increasingly apparent that these inhibitory or repulsive qualities of central nervous system (CNS) glia may be as important for neural pattern formation as its conducive function(s). Furthermore, inhibitory/repulsive properties of glia may play an essential role in the failure of CNS regeneration. Recent progress has been obtained in unraveling the molecular basis of inhibitory influences in the CNS. The present review discusses in detail tenascin-C and -R, two extracellular matrix glycoproteins which belong into a growing family of structurally related genes, are selectively expressed by glial lineages in the brain and exert both stimulatory and inhibitory/repulsive effects on CNS neurons.

KEY WORDS Adhesion, astrocyte, axon guidance, boundary, cell interactions, cytotactin, extracellular matrix, gradient, inhibition, janusin, J1-glycoproteins, neurite outgrowth, neuron-glia interactions neuron migration, neural development, oligodendrocyte, peripheral nerve, neuronal polarity, regeneration, restrictin, repulsion, segmentation, tenascin

INTRODUCTION

The development of the central nervous system of vertebrates unfolds as a finely tuned sequence of specificable steps. Several crucial events have been distinguished, for example the proliferation of epithelial stem cells, the migration of neuronal precursors from the ventricular zone to the cortex and to distinct nuclei, the construction of cortical layers, the establishment of the interneuronal circuitry, a wave of neuronal cell death and, eventually, the pruning and rearrangement of processes. These developmental stages are controlled by a variety of regulatory mechanisms which include specific intercellular, membrane-mediated interactions, the presence of limiting amounts of growth factors and synaptic activity (For review see Jacobson, 1991). It is well established that neuron-glia interactions plays a pivotal role in this context. For example, there is evidence that glial cells are critically involved in the guidance of migrating neuronal precursors and of extending neuronal fiber projections and in the segregation of forming neuronal aggregates and nuclei (Rakic, 1988, Steindler, 1993). Cell adhesion molecules (CAMs) and components of the extracellular matrix (ECM) are involved in the regulation of neuron-glia interactions. Most CAMs which have been described in the nervous system can be assigned to growing gene families, for example the cadherin superfamily for the calcium-dependent and the immunoglobulin superfamily for the calcium-independent adhesion mechanisms (Jessel, 1988, Geiger and Ayalon,

1992, Takeichi, 1991, Rathjen and Jessel, 1991, Schachner, 1991, Brümmendorf and Rathjen, 1993, Martini, 1994). Furthermore, ECM glycoproteins such as laminin, fibronectin and thrombospondin and their integrin-type receptors are expressed in developing neural tissues (Ruoslahti, 1988b, Sanes, 1989, Reichardt and Tomaselli, 1991, Hynes, 1992, Hynes and Lander, 1992, Letourneau *et al.*, 1994, Celio and Blümcke, 1994). Finally, proteoglycans which consist of a core glycoprotein and at least one covalently bound glycosaminoglycan chain (Hassel *et al.*, 1986, Ruoslahti, 1988a) have been detected both in neuronal and in glial cell lineages and may affect neurite growth in some cases (Herndon and Lander, 1990, Lander, 1993, Faissner *et al.*, 1994b, Letourneau *et al.* 1994, Faissner and Steindler, 1995).

On the other hand, it has recently become apparent that, in addition to conducive interactions, anti-adhesive, inhibitory and/or repulsive influences are of importance for the generation of ordered neural structures (Patterson, 1988, Keynes and Cook, 1992, Luo and Raper, 1994). Indeed, membrane-bound inhibitory components have been described which are implicated in development, de- and regeneration phenomena. These comprise the oligodendrocyte-derived inhibitors of neurite growth NI-35/250 (Caroni and Schwab, 1988a,b, Schwab and Caroni, 1988, Fawcett *et al.*, 1989, Bandtlow *et al.*, 1990, 1993, Bastmeyer *et al.*, 1991, Schwab *et al.*, 1993); and growth cone collapse-inducing components from the posterior somite (Keynes and Stern, 1984, Keynes and Lumsden, 1990, Davies *et al.*, 1990) and posterior chicken tectum (Walter *et al.*, 1987a,b, Cox *et al.*, 1990, Stahl *et al.*, 1990, Bastmeyer and Stuermer, 1993). These glycoproteins have in common that they regulate fiber growth by the induction of growth cone collapse, a phenomenon originally discovered when peripheral dorsal root ganglion (DRG) neurites were confronted with retinal neurites *in vitro* (Kapfhammer and Raper, 1987a,b). Recently, the 100 kD collapse-inducing molecule collapsin was obtained from chicken CNS (Luo *et al.*, 1993) and subsequent analysis showed that it bears homologies to fasciclin IV (fasIV) from Drosophila (Kolodkin *et al.*, 1992). Components highly related to fasIV have been identified in Drosophila and human and what appears to represent a novel gene family has been renamed semaphorins (Kolodkin *et al.*, 1993, Luo and Raper, 1994, Dodd and Schuchardt, 1995). Interestingly, in some cases proteoglycans are also thought to inhibit, rather than to promote, neurite growth (Snow *et al.*, 1990a,b, Oohira *et al.*, 1991, Snow and Letourneau, 1992, Friedlander *et al.*, 1994). In this context, the tenascin gene family, a growing group of structurally related ECM glycoproteins, is of particular interest, because tenascin gene products are expressed in developing neural tissues and they seem to embody both inhibitory and stimulatory properties for several neural cell types. These molecules might, therefore, serve as models to study both growth and avoidance behaviours of growing axons and exemplify the dual aspect of astroglial effects on developing neural tissues (Mason, 1993, Steindler, 1993, Faissner and Steindler, 1995).

STRUCTURE OF TN-C

The tenascin-gene family comprises the three distinct genes tenascin/cytotactin (also named tenascin-C, in the following referred to as TN-C), janusin/restrictin (also designated tenascin-R, subsequently TN-R) and tenascin-X (TN-X, for further review see Jones and Copertino, this volume and Erickson and Bourdon, 1989, Chiquet-Ehrismann, 1991, Erickson, 1993a,b, Tucker, 1994). TN-C glycoproteins of the ECM (Chiquet-Ehrismann *et al.*, 1986) have been independently discovered by several groups and named myotendinous antigen (Chiquet and Fambrough, 1984a,b), human glioma-mesenchymal extracellular matrix antigen (GMEM, Bourdon *et al.*, 1983, 1985), hexabrachion (Erickson and Inglesias, 1984, Erickson and Taylor, 1987), cytotactin (Grumet *et al.*, 1985), J1-200/220 (Kruse *et al.*, 1985, Faissner *et al.*, 1988), J1/tenascin (Faissner and Kruse, 1990) and polypeptide 150/225 (Gulcher *et al.*, 1986). TN-C is transiently expressed in developing organs such as the CNS (Kruse *et al.*, 1985, Grumet *et al.*, 1985, Gulcher *et al.*, 1986), the kidney (Aufderheide *et al.*, 1987), the intestine (Aufderheide and Ekblom, 1988) and the mammary gland (Chiquet-Ehrismann *et al.*, 1986), to name just a few (reviewed in Erickson and Bourdon, 1989). In most of these cases, TN-C is confined to circumscribed areas in a site-restricted fashion, for example in zones of epithelial-mesenchymal induction (Crossin *et al.*, 1986). Primary derived amino acid sequences have been reported for several species, e.g. for chicken (Spring *et al.*, 1989, Jones *et al.*, 1989), mouse (Weller *et al.*, 1991, Saga *et al.*, 1991), and human TN-C (Gulcher *et al.*, 1989, Siri *et al.*, 1991). The glycoproteins comprise a cysteine-rich segment at the amino terminus, subsequent $13\frac{1}{2}$ EGF-type repeats in the chicken and $14\frac{1}{2}$ EGF-type repeats in mouse and human followed by eight fibronectin type III (FNIII) homologous repeats and homologies to fibrinogen-β and -γ at the carboxyl terminus. Several isoforms have been described which contain alternatively spliced FNIII repeats between the fifth and the sixth FNIII repeat of the basic structure. The interspecies homology between analogous FNIII repeats reaches 90% while the intraspecies homology between FNIII modules is around 40%. Theoretically, a wealth of TN-C isoforms could potentially be encoded by various combinations of alternatively spliced FNIII repeats, but only a restricted number has hitherto been described in the mouse (Figure 3.1). Biochemically, TN-C glycoproteins display apparent molecular weights ranging from 190 kD to 320 kD and are linked to multimers by disulfide-bridges at the amino terminal end (Chiquet-Ehrismann *et al.*, 1986, Aufderheide *et al.*, 1987, Taylor *et al.*, 1989, Faissner and Kruse, 1990). The prevalent configuration of native TN-C is a hexamer which appears as six-armed structure, a so called hexabrachion, upon rotary shadowing and electron microscopy (Erickson and Inglesias, 1984, Vaughan *et al.*, 1987). The apparent molecular mass of the hexabrachion is about 1000 kD and it has a sedimentation constant of 6 s in glycerol gradients (Taylor *et al.*, 1989). TN-C glycoproteins from CNS tissues express

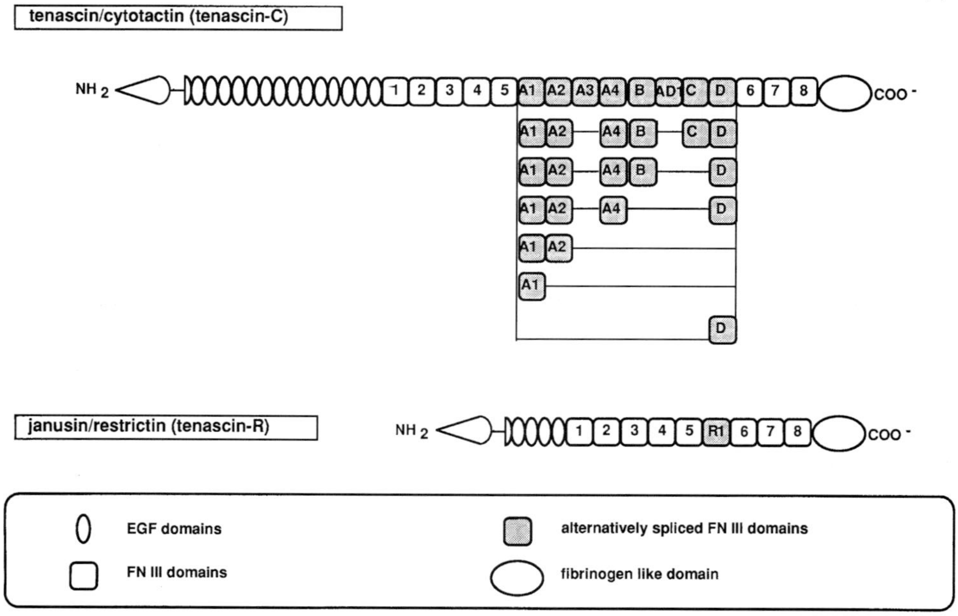

Figure 3.1 Structure of TN-C and TN-R. The drawing shows the structure of the two members of the tenascin gene family which are expressed in neural tissues. The largest known variant of human TN-C is depicted. The shaded boxes correspond to alternatively spliced FNIII repeats which are inserted between the 5th and 6th FNIII-motive of the smallest variant. The various combinations of FNIII repeats aligned below the large human variant represent those alternatively spliced isoforms which have hitherto been deciphered in the mouse. Theoretically, a large repertoire of combinations is conceivable. To which extent this potential is realized *in vivo* is presently unknown. Note that neither TNfnA3 nor TNfnAD1 have as yet been detected in mouse. TN-R is the second tenascin gene family component that has so far been isolated from neural tissues. Rat TN-R displays 70% homology to mouse TN-C and the FNIII domain R1 could constitute an alternatively spliced insert. The meaning of the geometrical symbols is explained in the legend, for further details see text and the chapter by Jones, this volume.

the L2/HNK-1 carbohydrate epitope, in contrast to the mesenchymally-derived glycoprotein (Kruse *et al.*, 1984, 1985, Chou *et al.*, 1986, Hoffman *et al.*, 1988, Probstmeier *et al.*, 1990).

TN-C EXPRESSION IN THE DEVELOPING CNS

Early expression of TN-C. In the chick embryo, TN-C is first expressed during gastrulation and appears in the basement membrane of the developing neural tube and notochord in cephalocaudal sequence (Crossin *et al.*, 1986). At later stages, TN-C is detectable in distinct areas of the developing CNS where its expression correlates with key events of neurohistogenesis such as neuronal migration, segregation of neuronal assemblies and extension of neuronal processes (Crossin *et al.*, 1986, Bartsch *et al.*, 1992a,b). Although TN-C-immunoreactivity has also been detected on the surface of neurites, most neuronal cell types do not express the glycoprotein by themselves. *In situ* hybridization studies have clearly established astrocytes as the major source of TN-C in the developing brain (Prieto *et al.*, 1990, Tucker, 1991, 1993, Bartsch *et al.*, 1992a, Tucker *et al.*, 1993, Mitrovic *et al.*, 1994). In addition, TN-C transcripts have been documented in the ependymal layer of postnatal mice and in a subpopulation of avian retinal neurons (Tucker, 1991). It has to be noted here, however, that most of the studies published so far do not distinguish individual isoforms of TN-C in this context. Interestingly, correlation of high Mr-isoform expression, that is of TN-C variants containing alternatively spliced FNIII repeats, with significant histogenetic events, has been pointed out for the developing chicken cornea where the immigration of neural-crest-derived prospective corneal epithelia correlates with the upregulation of the 230 kD isoform of chicken TN-C. Likewise, high Mr isoforms of TN-C prevail during the period of granule cell migration in the developing cerebellum, while the adult organ contains low levels of the smallest variant (Bartsch *et al.*, 1992a). Initial investigations using probes specific for the alternatively spliced FNIII domains have suggested a correlation between the expression of the high Mr variant of TN-C and processes of cell proliferation, cell migration and fiber outgrowth (Tucker *et al.*, 1993).

TN-C boundaries. Another remarkable feature of TN-C expression is its boundary-like expression in some regions of the developing CNS. It had been noted that prospective functional neurophysiological processing units are transiently demarcated by glial cells during cortical development. These glial boundaries could be visualized with the lectin PNA whose binding site displays a locally restricted expression, for example in the somatosensory barrel field (Steindler and Cooper, 1987, Cooper and Steindler, 1986a,b, Steindler, 1993). Recently, the observation that TN-C-immunoreactivity co-distributes with the lectin in several cases and transiently delineates barrels during the period of thalamocortical afferent ingrowth has advanced our knowledge about the molecular composition of PNA-positive glial boundaries (Figure 3.2, Crossin *et al.*, 1989, 1990, Steindler *et al.*, 1989a,b, 1990). TN-C by itself, however, does not express a PNA-binding site, while the cytotactin binding proteoglycan CTBP, which co-localizes with TN-C in the barrel field, is recognized by the lectin, which indicates molecular complexity of the boundary structure (Crossin *et al.*, 1989). Discrete distribution of TN-C glycoproteins has also been detected in other areas of both the developing PNS (Martini and Schachner, 1991, Wehrle-Haller *et al.*, 1991) and CNS. For example, TN-C depositions follow the subdivision between patch and matrix compartment of the forming striosome or of fiber tracts in the internal capsule,

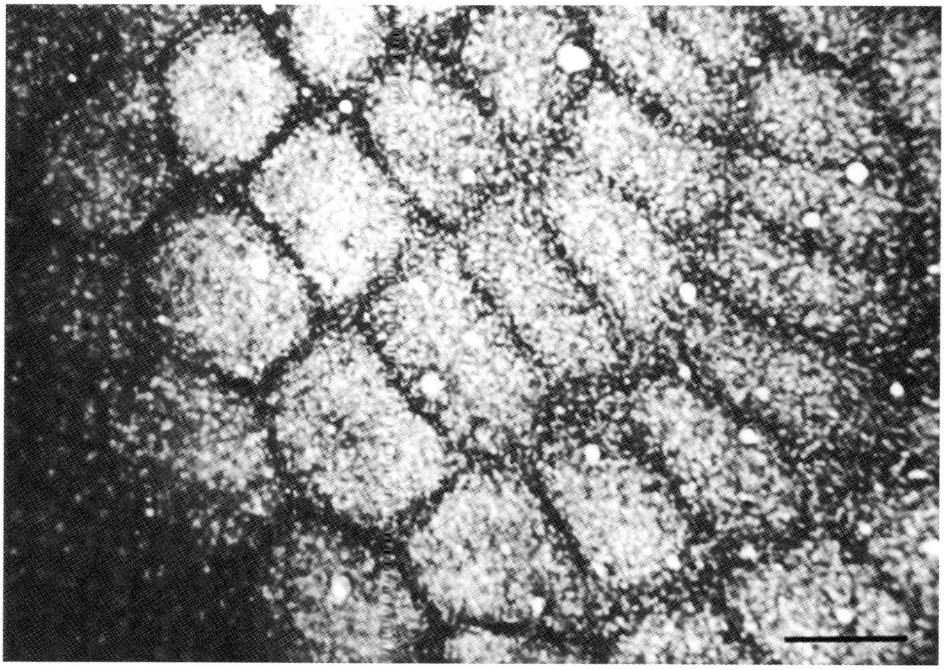

Figure 3.2 TN-C boundaries in the postnatal somatosensory cortex. The figure displays a top view on a flattened postnatal day 6 (P6) mouse somatosensory cortex which has been treated with polyclonal TN-C and developed with peroxidase-derivatized secondary antibodies. During that period, TN-C immunoreactivity delineates the outlines of prospective barrels, processing units for sensory information conveyed from the peripheral whisker pad sensory organs. The expression of TN-C is transient and down-regulated after thalamocortical afferent innervation has been accomplished. For further details see text and recent reviews (Steindler *et al.*, 1990, Steindler, 1993). Scale bar: 200 µm, photograph by courtesy of Dr. Steindler, Memphis, TN, USA.

separate functional compartments in nuclei of the brainstem, eg. the trigeminal nucleus, surround glomeruli of the the mouse olfactory bulb and subdivide motor centres in the vocalization control nuclei of the zebra finch songbird (Steindler *et al.*, 1990, Laywell and Steindler, 1991, O'Brien *et al.*, 1992, Steindler, 1993, Gonzalez *et al.*, 1993, Gonzalez and Silver, 1994, Burkhardt-Holm *et al.*, 1995). In this regard, it is noteworthy that expression of TN-C-like immunoreactivity has also been found in glial cells enveloping the forming olfactory glomeruli of the moth *Manduca sexta*, which might indicate a phylogenetic conservation of TN-C boundaries (Tolbert and Oland, 1990, Krull *et al.*, 1994a). Interestingly, a persisting boundary of TN-C-immunoreactivity has been revealed in the optic nerve head. This structure is believed to prevent immigration of oligodendrocyte precursors into the non-myelinated retina (Small *et al.*, 1987, Kiernan and ffrench-Constant, 1993, Bartsch *et al.*, 1994). On the other hand, homogeneous distribution of TN-C in zones of extensive fiber growth has also been documented, for example in the developing mouse optic nerve (Bartsch *et al.*, 1992b), the maturing postnatal mouse cerebellar cortex (Bartsch *et al.*, 1992a), the postnatal mouse cortex (Godfraind *et al.*, 1988, Jhaveri *et al.*, 1991, Sheppard *et al.*, 1991, Mitrovic *et al.*, 1994) and the regenerating sciatic nerve (Daniloff *et al.*, 1989, Martini *et al.*, 1990). These results imply that the growth of axon tracts through TN-C-containing territories is not generally impeded. In most of these cases, the expression of TN-C is transient and the glycoproteins are virtually undetectable in the majority of mature CNS structures. Down-regulation of TN-C message follows a well orchestrated temporal schedule. Thus, for example, TN-C mRNA declines in the mouse cortex between postnatal day 1 (P1) to P15, with an earlier down-regulation in layers 5 and 6 as compared to layers 1 and 2/3 (Mitrovic *et al.*, 1994). Notable exceptions are the olfactory bulb (Miragall *et al.*, 1990, Gonzalez *et al.*, 1993, Gonzalez and Silver, 1994) and a discrete band in the subventricular zones of zebra finch songbird (Burkhardt-Holm *et al.*, 1995) and mouse telencephali (Gates *et al.*, 1995). Interestingly, these territories are characterized by a certain degree of plasticity up to adult stages.

Regulation of TN-C expression. The factors which regulate the timing and the topography of TN-C expression are only partially known. The TN-C gene promoter contains a functional binding site for the homeobox transcription factor Evx-1, an AP-1/(TRE) site which may be activated by cellular signal transduction events (Jones *et al.*, 1990, 1992, Edelman and Jones, 1992), an upstream repressor or silencer region (Copertino *et al.*, 1995, Gherzi *et al.*, 1995), and a response element which is activated by mechanical stress in contracting collagen gels (Chiquet-Ehrismann *et al.*, 1994). Several growth factors have been reported to stimulate TN-C expression, for example TGF-β1, NGF and angiotensin II which enhance TN-C expression in embryonic fibroblasts, C6 glioma and vascular smooth muscle cells, respectively (Pearson *et al.*, 1988, Yavin *et al.*, 1991, Sharifi *et al.*, 1992). Likewise, interleukin-1, interleukin-4 and TNF-α enhance TN-C expression in some cell types (Rettig et al., 1994). Recently, it has been reported that bFGF, a growth factor with neurotrophic functions for PNS and CNS neurons (Unsicker *et al.*, 1987), elevates TN-C protein in astrocyte monolayer cultures (Meiners *et al.*, 1993). In the somatosensory barrel field, the appearance of TN-C boundaries is subsequent to the ingrowth of thalamocortical projections. The hollowing out of TN-C-immunoreactivity in the barrel area might indicate that TN-C boundaries in this area are sculptured by afferent fiber systems (Jhaveri *et al.*, 1991). This conclusion is supported by a recent study which demonstrates that TN-C message and protein are homogeneously distributed in the postnatal cortex after birth and both down-regulated in the prospective barrel hollows of layer 4, concomittant with the ingrowth of thalamocortical afferents during P3/P4, while neighbouring areas of the same layer and layers 1, 2/3, 5 and 6 still exhibit homogeneous distribution of TN-C (Mitrovic *et al.*, 1994). The dependence of boundary formation on neuronal afferents is further emphasized by reports that de-afferentation during the critical period can alter TN-C boundary representations in the somatosensory cortex, in the striosome and around axonal glomeruli in the olfactory bulb (Cooper and Steindler, 1989, O´Brien *et al.*, 1992, Gonzalez *et al.*, 1993, Gonzalez and Silver, 1994). Consistent with this view, immigrating retinal neurites reduce TN-C immunoreactivity on the surface of co-cultured astrocyte monolayers *in vitro* (Ard *et al.*, 1993). The neuronal factor(s) presumably responsible for the establishment of TN-C patterns and the role of neuronal activity in this context are as yet unknown. In this regard, it is noteworthy that glucocorticoids supress the expression of the glycoprotein in bone marrow stromal cells and in fibroblasts (Ekblom *et al.*, 1993).

ANTI-ADHESIVE PROPERTIES OF TN-C FOR NEURAL CELLS

The distribution of TN-C in boundary-like structures which separate anatomically defined, functional processing units of the nervous system could reflect that the glycoproteins serve to segregate assembling neuronal modules from each other in these areas. Several *in vitro* studies support this hypothesis. Thus, purified TN-C glycoproteins are a poor substrate for neural cell culture. For example, E10 chicken neurons or glial cells fail to attach to TN-C when plated under normal gravity conditions (gravity assay, 1xG, Friedlander *et al.*, 1988) and E11 chicken sympathetic, E14 rat mesencephalic, E18 rat hippocampal or postnatal day 6 (P6) mouse cerebellar neurons and E14 mouse cerebral astrocytes do not form monolayer cultures when the cells are plated on TN-C substrates and maintained for 24 hours *in vitro* (Friedlander *et al.*, 1988, Wehrle and Chiquet, 1990, Faissner and Kruse, 1990). Rather than providing a favorable environment for neural cells, TN-C displays repulsive properties. The combination of TN-C with poly-ornithine induces neuronal aggregate formation and neurite fasciculation under high density plating conditions of several types of embryonic and postnatal CNS neurons (Faissner and Kruse, 1990). Along these lines, purified TN-C causes fasciculation and reduces the branching of neurites leaving E3 chicken spinal cord explants (Wehrle and Chiquet, 1990). Likewise, the adhesion of the neural crest-derived neuronal cell lines N2A and PC12 to collagens I-V is inhibited when TN-C is bound to the substrate prior to cell addition (Bachmann *et al.*, 1995). The repulsive properties of TN-C are also apparent when neural cells are exposed to patterned substrates composed of supportive ECM components or a polycation and TN-C (Figure 3.3). Indeed, embryonic and postnatal CNS neurons preferentially grow on TN-C-free parts of patterned TN-C/polyornithine coverslips (Faissner and Kruse, 1990). Likewise, neurites growing out from embryonic chicken DRG explants avoid TN-C-containing areas of

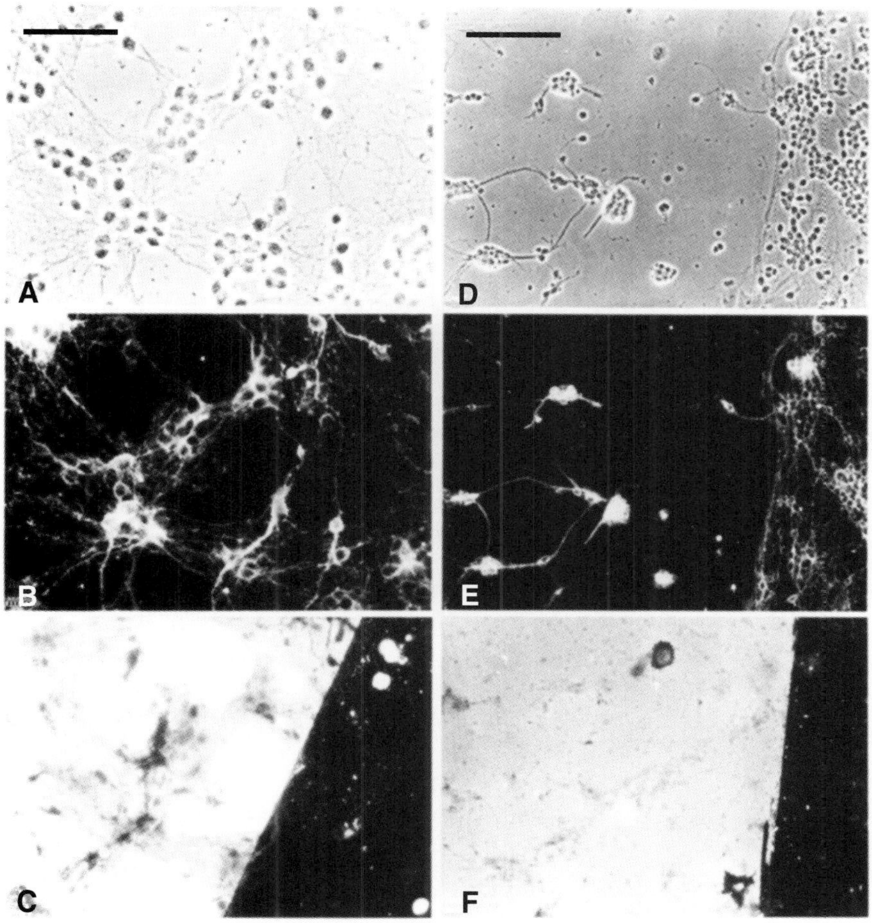

Figure 3.3 Repulsive properties of TN-C glycoproteins. Fibronectin glycoproteins (A–C) or TN-C purified from mouse brain (D–F) were exposed as patterned substrates in combination with poly-DL-ornithine and used for culture of mouse P6 cerebellar neurons for 3 days *in vitro*. After the culture period the neurons were stained with a monoclonal antibody to L1 (clone 324) followed by anti-rat TRITC (B, E) and counterstained with polyclonal fibronectin (C) or TN-C (F) antibodies and anti-rabbit FITC (F). Note that on patterned substrates neurons and their processes distribute to the TN-C-free areas of the coverslip (D, E), as confirmed by immunofluorescence labelling with specific antibodies (F), while fibronectin (A–C) does not exhibit repulsive properties. For details see Faissner and Kruse (1990); bars in (A) 40 μm (A–C), in (D) 100 μm (D–F).

patterned TN-C/laminin or -/fibronectin substrates (Crossin *et al.*, 1990, Taylor *et al.*, 1993). The repulsive properties of TN-C *in vitro* are consistent with the interpretation that these glycoproteins conceal neuronal assemblies and extending axonal pathways *in situ*. Repulsive effects of TN-C have also been observed in neuron-glia co-cultures where so-called *rocky* astrocytes produce circumscribed, locally enriched patches of TN-C which deflect neurons and neurites (Grierson *et al.*, 1990, Meiners *et al.*, 1993). TN-C might embody repulsive properties because it interferes with cell-spreading. Anti-spreading effects have been reported for both neural and non-neural cells, for instance rat or mouse embryonic day 18 (E18) hippocampal neurons (Lochter *et al.*, 1991), mouse E11 fibroblast and mouse L929 cells (Chiquet-Ehrismann *et al.*, 1988), neural crest cells (Tan *et al.*, 1987, Mackie *et al.*, 1988, Epperlein, 1988), mesodermal cells from gastrula stage embryos (Riou *et al.*, 1990) and human melanoma cells (Halfter *et al.*, 1989). The anti-spreading properties of TN-C are also observed when TN-C is combined with fibronectin, laminin and polycations and is clearly documented when the cells are grown in the presence of TN-C added as soluble component to the culture medium rather than as component to the culture substrate (Tan *et al.*, 1987, Mackie *et al.*, 1988, Chiquet-Ehrismann *et al.*, 1988, Riou *et al.*, 1990, Faissner and Kruse, 1990, Lochter *et al.*, 1991). In this context, reduced cellular spreading after plating might be a consequence of the interference with focal contact point formation, a phenomenon observed when TN-C was added in soluble form to cultured bovine aortic endothelial cells (Murphy-Ullrich *et al.*, 1991).

DISTINCT SITES IN TN-C MEDIATE NEURAL CELL-BINDING AND REPULSION

Cell-binding sites in TN-C. The anti-spreading and repulsive properties of TN-C were surprising in view of the fact that the glycoproteins had originally been found to mediate attachment of CNS neurons to astrocyte monolayers (Kruse *et al.*, 1985, Grumet *et al.* 1985, Faissner *et al.*, 1988). This conclusion had been derived from short-term adhesion assays (30 minutes at room temperature, Keilhauer *et al.*, 1985) performed in the presence of polyclonal antibodies to TN-C or to control proteins. This would suggest cell-binding site(s) in the molecule, a conclusion consistent with results obtained when various cell types were exposed to the purified molecule in binding studies. Indeed, erythrocytes, P6 mouse cerebellar and E9 chicken retina neurons, E10 chicken embryo fibroblasts, mouse L929 and U251MG human glioma cells attach to TN-C in short-term assays (Chiquet-Ehrismann, 1986, Friedlander *et al.*, 1988, Bourdon and Ruoslahti, 1989, Nörenberg *et al.*, 1992). Studies with proteolytic fragments suggested a cell-binding site in 35 kD cyanogen bromide and a 60 kD pronase fragments which comprise the carboxyl terminal end of TN-C monomers (Friedlander *et al.*, 1988, Chiquet *et al.*, 1991). This conclusion has been confirmed with recombinant proteins representing the carboxyterminal globule of human TN-C and the integrin $\alpha 2\beta 1$ has been inferred as a potential receptor of this interaction (Aukhil *et al.*, 1993, Sriramarao *et al.*, 1993, Joshi *et al.*, 1993). Interestingly, TN-C binds heparin (Marton *et al.*, 1989, Faissner *et al.*, 1990, Aukhil *et al.*, 1993, Weber *et al.*, 1995, Fischer *et al.*, 1995) and the higher affinity binding site for this glycosaminoglycan is located in the fibrinogen-like domain (Aukhil *et al.*, 1993). This opens the possibility that a heparan sulfate proteoglycan, possibly syndecan, participates in binding cells to this part of TN-C (Figure 3.4, Salmivirta *et al.*, 1991). This conclusion is compatible with the finding that the binding interaction does not require divalent cations and is sensitive to the presence of soluble heparin in the assay medium, pre-treatment of cells with heparitinase and undersulfation of cell surface proteoglycans (Joshi *et al.*, 1993). Glial cells bind rather well to the carboxyterminal fibrinogen-like domain of TN-C, while CNS neurons showed weak attachment to this part of the glycoprotein (Prieto *et al.*, 1992). A second cell binding site has been attributed to the third

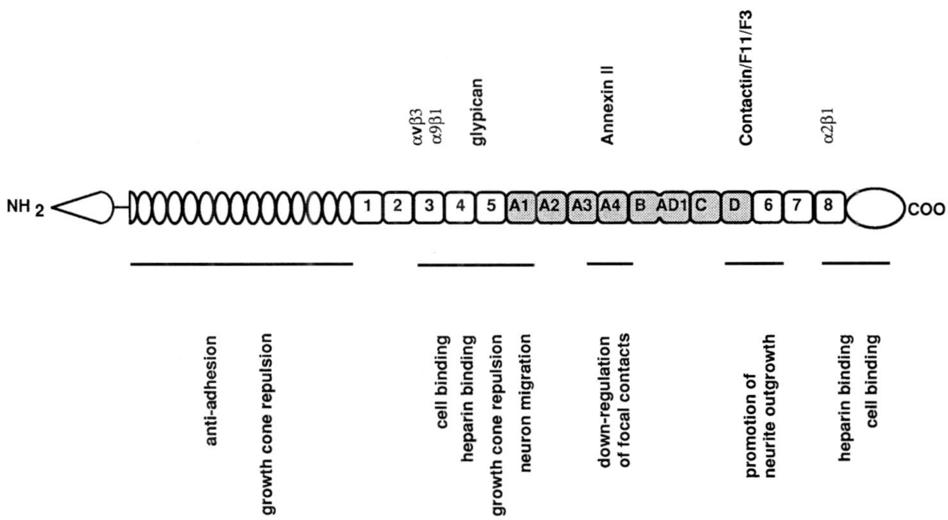

Figure 3.4 Localisation of functional domains of TN-C. The figure condenses the results of several studies on structure-function interrelationships of TN-C glycoproteins. The functional domains are tentatively projected onto the large human isoform. The map is based on *in vitro* perturbation assays using monoclonal antibodies with known binding sites and on the application of proteolytic fragments or recombinant proteins produced in bacterial expression systems in functional assays. It has to be pointed out that, with the exception of the RGD-motive in TNfn3, the primary amino acid sequences of the functional sites have not yet been determined. Proposed TN-C receptors are designated above their assumed binding regions, functional attributions are outlined below the TN-C structure. For the meaning of the geometrical symbols see Figure 3.1 and for further details see text.

FNIII repeat of the basic structure, that is TNfn3 (Figure 3.4). $\alpha v\beta 3$ mediates binding of several cell types to this domain in an RGD-dependent manner (Joshi *et al.*, 1993, Sriramarao *et al.*, 1993, Aukhil *et al.*, 1993, Prieto *et al.*, 1993). Recently, a second receptor for this domain, $\alpha 9\beta 1$, has been revealed. This integrin does, however, not require the RGD sequence for cell attachment to TNfn3 (Yokosaki *et al.*, 1994). The role of glycolipid ligands of TN-C in the context of cell binding can presently not be evaluated (Crossin and Edelman, 1992).

Anti-adhesive sites in TN-C. The fact that TN-C contains defined cell attachment sites and functional binding motifs for integrins implies that the overall anti-adhesive and repulsive properties of the glycoprotein are effected by additional, separate domains which, eventually, override the conducive properties of the molecule (Faissner, 1993). This interpretation has been confirmed by the isolation of TN-C fragments which embody anti-adhesive properties. Thus, anti-adhesive activities have been attributed to the EGF-type repeats (Spring *et al.*, 1989, Prieto *et al.*, 1992), to the alternatively spliced domains TNfnA-D of human tenascin (Murphy-Ullrich *et al.*, 1991), and to the carboxy-terminal end encompassing TNfn7,8 (Prieto *et al.*, 1992, Chiquet *et al.*, 1991), areas clearly distinct from the cell-binding sites described above (Figure 3.4). An important implication of the studies with isolated TN-C domains is that anti-adhesiveness apparently does not require multivalency of the ligand. Another implication of the multimodular, multifunctional organisation is that cell-TN-C interactions probably evolve as a sequence of several distinct steps. The latter might begin with an initial cell binding event and proceed to recognition of defined motifs in the anti-adhesive domains, which might trigger downstream processes which eventually interfere with cytoskeletal rearrangements necessary for the spreading process to occur. This two-phase model is supported by the determination of physical cell-TN-C interaction forces in a quantifiable binding assay. For example, the initial forces generated between U251MG glioma cells and fibronectin are comparable to those measurable between these cells and TN-C (10^{-5} dyne). Thereafter, spreading leads to an 125-fold increase of cell-substrate adhesion forces on fibronectin while cells remain rounded on TN-C, which prevents the increase of substrate adhesion (Lotz *et al.*, 1989). Interestingly, it had been noted in earlier studies that the anti-spreading effect can be abolished by blocking the cell binding site in the carboxy-terminal part of TN-C with mAb Tn68 (Chiquet-Ehrismann *et al.*, 1988). This is consistent with the idea that cell-binding would precede the generation of the repulsive effect in a sequential process. This model implies receptor cross-talk, a concept which has already been discussed for integrins (Hynes, 1992). The receptors for the assumed anti-adhesive signal(s) are presently unknown, as are the nature of the inferred downstream processes which presumably involve 2nd-messenger pathways (Igarashi *et al.*, 1993, Doherty and Walsh, 1994). In this regard, it is an exciting discovery that phosphotyrosine phosphatases (PTPs) have been shown to interact with TN-

C. Thus, the chondroitin sulfate proteoglycans phosphacan and PTP-β, which are both expressed in the CNS, are TN-C ligands (Grumet *et al.*, 1994, Barnea *et al.*, 1994, Maurel *et al.*, 1994). This opens an avenue to the intersection of TN-C-cell-interactions with growth-factor controlled phosphorylation pathways, a link which has recently been discussed for the adhesion molecule L1 in relation to the bFGF receptor (Fischer *et al.*, 1991, Williams *et al.*, 1994a,b). It is noteworthy in this regard that tyrosine phosphorylation of tubulin subunits has been implicated in growth cone movement and can be affected by the cell adhesion molecule L1 (Atashi *et al.*, 1992, Bixby and Harris, 1991). Whether the TN-C chondroitin sulfate proteoglycan ligand CTBP (Hoffman and Edelman, 1987, Hoffman *et al.*, 1988) also belongs into the phosphatase family is currently unknown. In view of the observation that the alternatively spliced FNIII domains TNfnA-D induce down-regulation of focal contact points (Murphy-Ullrich *et al.*, 1991), it would be interesting to learn whether the above mentioned proteoglycans bind to this region of TN-C. In this context, annexin II has recently been proposed as specific TN-C receptor which binds to the alternatively spliced part of the glycoprotein. Whether this 35 kD protein mediates the anti-spreading properties of TN-C for endothelial and U251MG glioma cells is, however, still an open question (Chung and Erickson, 1994). Likewise, the receptor(s) which transduce the inhibition of fibroblast proliferation stimulated by TN-C are currently unknown (Crossin, 1991).

TN-C AND NEURAL CELL MIGRATION

The conjunction of cell-binding and anti-adhesive sites in TN-C endows the glycoprotein with qualities which are ideally suited to modulate adhesion and de-adhesion phenomena (Chiquet-Ehrismann, 1991, Sage and Bornstein, 1991). The dichotomy of bond formation and subsequent release is a hallmark of cell migration processes. Therefore, it was expected that effects of TN-C on cellular movement and/or motility in several systems would be observed. Thus, TN-C glycoproteins are thought to regulate migration of mesodermal cells in amphibian gastrulation and of dispersing neural crest cells and their derivatives, e.g. corneal epithelia (Tan *et al.*, 1987, Mackie *et al.*, 1988, Bronner-Fraser, 1988, Stern *et al.*, 1989, Riou *et al.*, 1990, Kaplony *et al.*, 1991). With regard to CNS tissue, antibody perturbation assays performed with postnatal cerebellar explants demonstrated that TN-C is involved in the translocation of granule cell neurons from the external to the internal granule cell layer, a pivotal process of postnatal cerebellar development (Chuong *et al.*, 1987, Bartsch *et al.*, 1992a, Husmann *et al.*, 1992). In this case, a region presumably implied in the control of neuron migration could be assigned to FNIII repeats 3–5 proximal of the alternative splice site using a panel of specific mAbs (Figure 3.4, Husmann *et al.*, 1992). Whether this regulatory function is linked to the cell binding site in TNfn3 (see above) is currently not known. Also in the E10–E13 mouse mesencephalon, the glycoprotein has been detected on the surface of radial glial processes which are thought to guide the ventral migration of dopaminergic neurons (Kawano et al., 1995). Interestingly, TN-C exhibits dual, opposite effects with respect to cell migration. Indeed, when added as soluble component to the culture medium, TN-C reduces migration rates of neuroblastoma cells (Halfter *et al.*, 1989) *in vitro*, perhaps because of interference with cytoskeletal organization.

Inhibitory properties of TN-C for cell migration could also be substantiated for oligodendrocyte progenitors, the so-called O2A-cells. O2A cells are believed to represent the precursors of the myelin-forming oligodendrocytes of the mature CNS. The progenitors are generated in distinct ventricular zones of the CNS and emigrate from there into the territories they are destined to myelinate. This process has been carefully studied in the developing optic nerve of the mouse where the first progenitors are detectable around the perinatal period (reviewed by Kiernan and ffrench-Constant, 1993). Interestingly, later myelin formation ceases close to the optic fissure and the retina of most mammals is not myelinated. The confinement of the myelination territory correlates with a boundary of TN-C in the optic nerve head (Bartsch *et al.*, 1994). It could be demonstrated that purified TN-C drives O4-positive, O1-negative immature oligodendrocytes *in vitro* into TN-C-free parts of a patterned TN-C/poly-ornithine coverslip. Subsequent investigations have shown that purified TN-C in combination with poly-L-lysine reduces the motility of O2A-progenitors obtained from rat optic nerve *in vitro*, as assessed by time-lapse video-microscopy. Remarkably, this is not due to anti-adhesive properties of TN-C per se because the same substrate does not reduce motility of CG-4 cells, which are thought to constitute a cell line model of the O2A progenitor. These observations are consistent with the hypothesis that TN-C represents a stop signal for oligodendrocyte precursors in the optic nerve head (Kiernan and ffrench-Constant, 1993 and personal communication). The freezing of glial cell motility by TN-C glycoproteins is not restricted to the CNS. In fact, a recent report stresses that although TN-C promotes neurite outgrowth from embryonic chicken DRG (see below), the motility of satellite cells derived from this organ is inhibited (Wehrle-Haller and Chiquet, 1993).

TN-C AND THE CONTROL OF NEURITOGENESIS

Neural tissue explants. The establishment of a network of interneuronal connections is of prime importance for the structure and the function of neural tissues. Astrocyte scaffolds guide advancing growth cones and ECM components such as laminin and fibronectin are believed to play a role in this context (Sanes, 1989, Reichardt and Tomaselli, 1991, Bixby and Harris, 1991). In view of its assumed role in neuron-glia interactions, the influence of TN-C on neurite growth has been examined by several investigators (Table

3.1). It turned out that in addition to cell binding and repulsive sites for various cell types TN-C glycoproteins also possess neurite outgrowth promoting properties. For example, purified TN-C glycoproteins on glass promote neurite growth from E3 chicken spinal cord and a substrate comprising a mixture of TN-C with laminin enhances neurite growth from E7 to E8 chicken DRG (Wehrle and Chiquet, 1990, Taylor *et al.*, 1993). This response, however, is age-dependent because fiber outgrowth from E10 chicken DRG is inhibited when TN-C alone or combined with either fibronectin or laminin as adsorbed to plastic is used as culture substrate (Crossin *et al.*, 1990). In addition, lineage-dependence of the growth response is indicated by the observation that E6 chicken retina explants do not show any fiber growth on TN-C-containing substrates (Taylor et al, 1993, Perez and Halfter, 1993). These observations could reflect differential expression of TN-C receptors or linkage of these receptors to different downstream effector mechanisms affecting the growth cone. As far as DRG explants are concerned, neurite growth on TN-C can be blocked by antibodies to $\beta1$-integrin and by the addition of heparin to the culture medium (Wehrle-Haller and Chiquet, 1993). This would implicate the carboxyterminal domain, which shows homologies to fibrinogen-β and -γ, in the control of PNS neurite formation, because this region of TN-C harbours binding sites for heparin and $\alpha2\beta1$ integrin (see above, Aukhil *et al.*, 1993, Sriramarao *et al.*, 1993, Joshi *et al.*, 1993).

Dissociated neurons. Distinct responses to TN-C have also been observed when suspensions of singular neurons instead of neural tissue explants were cultured on TN-C-containing substrates (Table 3.1). For example, distinct classes of leech neurons exhibit increased neurite formation on TN-C-containing glycoprotein preparations adsorbed to plastic (Masuda-Nakagawa and Wiedemann, 1992). Because dissociated neurons from higher organisms do not develop monolayer cultures on plain TN-C (see above), the glycoprotein has been combined with supportive ECM constituents or polycations. Under these conditions, TN-C promotes neurite outgrowth by E3 chicken spinal cord, E6 chicken retinal ganglion, E8–E11 chicken sensory and E8 chicken sympathetic neurons (Wehrle and Chiquet, 1990, Perez and Halfter, 1993, Bartsch *et al.*, 1995), E14 mesencephalic and E18 hippocampal neurons from rat and mouse (Lochter *et al.*, 1991) and P6 mouse cerebellar granule cells (Husmann *et al.*, 1992). Growth stimulation does not appear to affect all fiber compartments in the same way. Rather, it seems that TN-C polarizes E18 hippocampal neurons and preferentially stimulates the longest as compared to the other processes of this cell type (Lochter and Schachner, 1993).

Perturbation studies with a selection of mAbs to distinct areas of TN-C showed, that neurite outgrowth promotion involves a region of the FNIII repeats around or in the carboxyterminal region distal to the splice site and distinct from domains contributing to other TN-C functions (Figure 3.4, Lochter *et al.*, 1991, see above). This attribution has recently been confirmed by experiments in which purified recombinant domains encompassing this region

TABLE 3.1 INFLUENCE OF TENASCIN GLYCOPROTEINS ON NEURITE OUTGROWTH

CULTURE SYSTEM	HOMOGENEOUS TN-C SUBSTRATES	PATTERNED TN-C SUBSTRATES	ADDITION OF SOLUBLE TN-C TO CULTURE MEDIUM	REFERENCES
TENASCIN-C				
E6 chicken retina explants	No fiber outgrowth on plain TN-C and TN-C/LN substrates	Fibers deflected from TN-C-containing parts of patterned TN-C/basal lamina and TN-C/LN substrates	Soluble TN-C leads to 50% reduction of fiber growth on collagen gels	Taylor *et al.*, 1993 Perez and Halfter, 1993
E3 chicken spinal cord explants	Acceleration of fiber growth and reduction of fiber branching on plain TN-C-coated glass			Wehrle and Chiquet, 1990
E7–E8 chicken DRG explants	Increased growth cone speed on TN-C/LN substrates	Growth cones deflected from TN-C-containing parts of patterned TN-C/LN substrates	Soluble TN-C does not affect neurite growth on LN	Taylor *et al.*, 1993
E4 chicken DRG explants	Stimulation of neurite out-growth by TN-C substrate on glass	Unhindered growth of neurites on patterned substrates of TN-C versus LN or FN on glass	Soluble TN-C does not affect neurite growth on LN or FN	Wehrle-Haller and Chiquet, 1993
E8/E9 chicken DRG ganglia	In contrast to E4, delayed neurite outgrowth at slower rate on homogeneous TN-C on glass	Similar to E4 ganglia	Similar to E4 ganglia	Wehrle-Haller and Chiquet, 1993

CULTURE SYSTEM	HOMOGENEOUS TN-C SUBSTRATES	PATTERNED TN-C SUBSTRATES	ADDITION OF SOLUBLE TN TO CULTURE MEDIUM	REFERENCES
E10 chicken DRG explants	Poor or no neurite outgrowth on plain TN-C, concentration-dependent reduction of neurite growth on TN-C/LN and TN-C/FN substrates	Fibers deflected from TN-C-containing parts of patterned TN-C/LN and TN-C/FN substrates	Soluble TN-C leads to reduced fiber growth on conducive substrates	Crossin *et al.*, 1990
Dissociated single cell culture of *Manduca sexta* antennal-lobe neurons	TN-C applied to LN/conA substrates on glass inhibits neurite outgrowth	Neurites of most neurons deflected at boundaries of TN-C alternating with LN/conA in parallel stripes	Soluble TN-C reduces neurite outgrowth on LN/ConA substrates	Krull *et al.*, 1994b
Isolated single Retzius (R) or anterior pagoda (AP) neurons from Leech	TN-C enriched extracts obtained from leech ganglion chains as coated on plastic promote neurite outgrowth			Masuda-Nakagawa and Wiedemann, 1992
Dissociated E6 chicken retinal ganglion cell culture	Promotion of neurite growth on TN-C/PORN when applied at 20 μg/ml; stimulation decreases to PORN control levels with increasing TN-C coating concentration			Bartsch *et al.*, 1995
Dissociated single cell culture of E18 rat hippocampal neurons	Preferential stimulation of longest neurite on TN-C/PORN substrates which results in enhanced neuron polarity			Lochter and Schachner, 1993

CULTURE SYSTEM	HOMOGENEOUS TN-C SUBSTRATES	PATTERNED TN-C SUBSTRATES	ADDITION OF SOLUBLE TN TO CULTURE MEDIUM	REFERENCES
Dissociated single cell culture of E17 rat hyothalamus		Cell bodies and processes deflected from TN-C-expressing "rocky" astrocyte surfaces		Grierson *et al.*, 1990
Dissociated single cell cultures of E7 chicken retina	Reduced neurite growth on TN-C containing NC-implants from CNS lesions			McKeon *et al.*, 1991
TENASCIN-R				
E6 chicken retina explants	No fiber outgrowth on plain TN-R and TN-R/LN substrates	Fibers deflected from TN-R-containing parts of patterned TN-R/LN substrates		Taylor *et al.*, 1993
E7-E8 chicken DRG exlants	Increased growth cone speed on TN-R/LN substrates	Growth cones deflected from TN-R-containing parts of patterned TN-R/LN substrates	Soluble TN-R does not Influence growth cone movement on LN	Taylor *et al.*, 1993
P7-8 mouse cerebellar microexplants		Deflection of outgrowing neurites by TN-R on patterned TN-R/PLL substrates		Taylor *et al.*, 1993
Dissociated suspensions of mouse cerebellar neurons	Nearly no growth of processes on plain TN-R and reduced neurite outgrowth on mixed TN-R/LN adsorbed to NC			Pesheva *et al.*, 1989 Pesheva *et al.*, 1993

CULTURE SYSTEM	HOMOGENEOUS TN-C SUBSTRATES	PATTERNED TN-C SUBSTRATES	ADDITION OF SOLUBLE TN TO CULTURE MEDIUM	REFERENCES
Single cell suspensions from embryonic chicken retina	Retinal neurons adhere to plain TN-R adsorbed to NC, but do not grow neurites. Unhindered neurite growth on mixed TN-R/LN substrates			Rathjen *et al.*, 1991
Single cell suspension of E18 rat hippocampal neurons	Neurite outgrowth and enhanced length of longest fiber on plain TN-R adsorbed to NC; enhanced neuronal polarity on a TN-R/PORN substrate.			Lochter *et al.*, 1993

The table summarizes the outcome of several studies devoted to the influence of tenascin glycoproteins on neurite growth. Abbreviations used in the table are TN-C: tenascin-C, TN-R: tenascin-R, LN: laminin, FN: fibronectin, PLL: poly-L-Lysine, PORN: poly-DL-Ornithine, NC: nitrocellulose. For details see text and the references indicated in the right column.

were used instead of the intact glycoprotein (Faissner *et al.*, 1995). Interestingly, the Ig-superfamily member contactin/F11/F3 (Ranscht and Dours, 1988, Gennarini *et al.*, 1989, 1991, Brümmendorf *et al.*, 1989) has been proposed as a neuronal TN-C receptor (Zisch *et al.*, 1992). This GPI-linked glycoprotein binds to a region in close vicinity of the distal, downstream alternative splice-site of TN-C, within the neurite outgrowth promoting region pinpointed with the help of mAb J1/tn2 and bacterially expressed FNIII domains (Lochter *et al.*, 1991, Vaughan *et al.*, 1994 Faissner *et al.*, 1995). It is, however, unclear whether contactin/F11/F3 conveys neurite outgrowth promoting properties of TN-C to migrating growth cones and, in fact, some pieces of evidence argue against this possibility. Most importantly, the glycoprotein is believed to interact specifically with TNfn5,6, a pair of FNIII repeat only present in the non-spliced TN-C variant. Insertion of alternatively spliced FNIII repeats disrupts the binding of the receptor to the glycoprotein (Vaughan *et al.*, 1994). Yet, the high molecular weight alternatively spliced TN-C variants prevail during the early stages of neural development, when the neuronal networks are being established (Crossin *et al.*, 1989, Faissner and Schachner, 1995). The integrin $\alpha 8\beta 1$ has been proposed as an alternative receptor for the neurite outgrowth promoting properties of TN-C. Antibodies to the subunits of this integrin receptor and to FNIII repeats TNfn6-8 inhibit neurite outgrowth promotion of embryonic chicken dorsal root ganglion and motoneurons by TN-C (Varnum-Finney *et al.*, 1995). Analysis of embryonic day 7 (E7) to E8 chicken DRG growth cones advancing in a TN-C/laminin as compared to a solely laminin-containing territory by time-lapse videomicroscopy revealed that increased neurite lenghts result from enhanced growth cone locomotion (Taylor *et al.*, 1993). TN-C promotes neurite outgrowth by most of the neurons studied so far. A notable exception, however, has been reported in the moth *Manduca sexta*, where TN-C in combination with the lectin conA, an efficient substrate for the growth and differentiation of insect neurons, completely blocks neurite outgrowth (Krull *et al.*, 1994b).

Repulsion of growth cones. Despite the overall growth promoting properties of TN-C, neurites are deflected from TN-C-containing areas in choice situations on patterned substrates (Table 3.1, Figure 3.3). This holds true for singular rat E14 mesencephalic, E18 hippocampal. and mouse P6 cerebellar neurites on patterned TN-C/poly-ornithine and individual E8 chicken DRG growth cones on patterned TN-C/laminin substrates (Faissner and Kruse, 1990, Taylor *et al.*, 1993). Likewise, fibers growing out from E10 chicken DRG explants avoid TN-C-containing areas of patterned TN-C/laminin or TN-C/fibronectin and E6 chicken retinal explants fail to enter TN-C territories on basal lamina substrates (Crossin *et al.*, 1990, Perez and Halfter, 1993). Yet, also in this context the lineage of the neuron seems important, because a subpopulation of neurons from moth olfactory glomeruli grows neurites which ignore TN-C stripes alternating with laminin/ConA on a nitrocellulose support, while the majority of fibers are efficiently deflected (Krull *et al.*, 1994b). Antibody perturbation assays suggest that repulsion of E8 chicken DRG growth cones could be caused by the EGF-type and by the FNIII-repeats 4–5 of TN-C (Taylor *et al.*, 1993). There are precedents in the literature that molecules which permit neurite growth as homogeneous substrates may inhibit fiber formation when exposed as pronounced gradient. For example, temporal retinal neurites advance on lawns of posterior tectal membranes, although the latter contain a growth cone collapse-inducing activity that deflects neurites in choice situations when posterior and anterior tectal membranes are applied in parallel stripes on the culture substrate (Walter *et al.*, 1987a,b, Stahl *et al.*, 1990). In an intermediate way of exposure, that is when the posterior tectal membranes are presented as homogeneous, yet graded substrate, the steepness of the gradient proves decisive (Baier and Bonhoeffer, 1992). Changes of more than 5% per 25 μm result in a reduction of neurite growth. In light of these findings, sharp TN-C boundaries *in vitro* constitute local 100% step gradients, a situation which is also realized in TN-C boundary areas *in vivo*.

TN-C IN NEURAL TISSUE LESIONS

The association of TN-C with development and its down-regulation after histogenesis in most organs is consistent with the hypothesis that the glycoprotein is associated with tissue remodelling (Ekblom and Aufderheide, 1989, Erickson and Bourdon, 1989). This raises that possibility that it might play a role in the context of tissue lesion and pathology, an idea which has meanwhile been confirmed by numerous studies in pathology laboratories (see Chapter IV, this volume). As far as neural tissues are concerned, an association of TN-C with neural lesion has first been demonstrated for the neuromuscular junction. After denervation, TN-C is up-regulated in the lesion area and stays at the original synaptic site until re-innervation has proceeded. It is an interesting possibility that TN-C might contribute to the known local specificity of the neuromuscular synapse, which contains specific cues directing and halting regrowing motor fibers at the original innervation site (Gatchalian *et al.*, 1989, Daniloff *et al.*, 1989, Irintchev *et al.*, 1993). TN-C also shows a dramatic up-regulation in the lesioned peripheral nerve, where its appearance and subsequent down-regulation closely follows the schedule of retrograde nerve fiber degeneration and subsequent regrowth to the peripheral target (Daniloff *et al.*, 1989, Martini *et al.*, 1990). In both cases, TN-C is found associated with the ECM, in particular with collagen fibers, and with Schwann cell basal laminae as deduced from immunoelectron microscopy. This co-distribution might reflect interactions of TN-C with ECM constituents, e.g. various collagens, fibronectin and heparin, which have been detected by *in vitro* binding assays (Chiquet-Ehrismann *et al.*, 1988, Marton *et al.*, 1989, Lightner and Erickson, 1990, Faissner *et al.*, 1990, Probstmeier *et al.*, 1990, Chiquet *et al.*, 1991, Martini, 1994).The functional significance of TN-C expression in the regenerating PNS has recently been highlighted by antibody perturbation studies. The continuous application of polyclonal antibodies to TN-C resulted in a quantifiable reduction in the degree of re-innervation of the denervated neuromuscular junctions both in frog and mouse *in vivo* (Mège *et al.*, 1992, Langenfeld-Oster *et al.*, 1994). This would imply a supportive role of TN-C for neurite growth in the PNS, an interpretation consistent with the promotion of DRG-neurites observed *in vitro* (see above, Table 3.1). This conclusion has obtained support from recent observations in the C57BL/Wld(s) mutant mice, where regeneration of efferent fibers could only be found in the TN-C-positive muscle branch, whereas the TN-C-negative cutaneous branch was devoid of growing axons after experimental lesion to the femoral nerve (Fruttiger *et al.*, 1995). Interestingly, the regenerating caudal spinal cord of the newt, a CNS structure, also exhibits a sustained elevation of TN-C protein and mRNA expression upon lesion (Caubit *et al.*, 1994). In contrast, TN-C upregulation in the mouse dorsal root entry zone after lesion of the afferents has been interpreted as forming an active barrier which prevents ingrowth of regenerating DRG-derived afferents into the dorsal horn of the spinal cord (Pindzola *et al.*, 1993). TN-C glycoproteins exhibit enhanced expression after lesion to CNS tissues (McKeon *et al.*, 1991, Laywell *et al.*, 1992), although to a much lower extent than in peripheral nerves, and membranes attached to nitrocellulose brain implants reduce neurite outgrowth from dissociated retinal neurons differentiating on this substrate (McKeon *et al.*, 1991). These inhibitory properties, however, could also be due to enhanced expression of proteoglycans in these preparations and a direct link of inhibitory activity to TN-C could not yet be established (McKeon *et al.*, 1991). Therefore, the biological significance of up-regulation of TN-C in CNS wounds, both in experimental animals, human subjects (Brodkey *et al.*, 1995) and in human CNS tumors (Higuchi *et al.*, 1993, Yoshida *et al.*, 1994, Zagzag *et al.*, 1995) remains at present elusive.

STRUCTURE AND DISTRIBUTION OF TENASCIN-R

TN-R, the second member of the tenascin gene family, has originally been discovered as lower Mr constituent of the J1-antigen, a group of four developmentally-regulated glycoproteins of forming neural tissues (Kruse *et al.*, 1985). The J1-antigen had been defined by polyclonal antibodies generated against a prominent 160 kD component of *rest-L2*, a fraction CNS glycoproteins expressing the L2/HNK-1 carbohydrate epitope. J1-antibodies react in Western blot experiments with four major components of 160 kD, 180 kD, 200 kD and 220 kD apparent Mr which display differential expression during CNS development. J1-200/220 corresponds to TN-C (see above) and is expressed during late embryonic and early postnatal stages of development. In contrast, mouse J1-160/180, also named janusin, shares some epitopes with TN-C, but represents a clearly distinct entity on its own which reaches higher expression levels after birth and persists into adulthood (Faissner *et al.*, 1988, Pesheva *et al.*, 1989). Subsequent molecular genetic studies have shown that J1-160/180 (janusin) exhibits substantial structural resemblances to TN-C (Figure 3.1). Thus, a cysteine-rich aminoterminus is followed by $4\frac{1}{2}$ EGF-type and nine FNIII repeats and the sequence is completed by homologies to fibrinogen β and γ. The homology to TN-C FNIII repeats is about 70% at corresponding positions, underscoring the relatedness of both glycoproteins (Pesheva *et al.*, 1989, Fuss *et al.*, 1993). A similar molecule which also expresses the L2/HNK-1 epitope has recently been described in the chick and named restrictin (Rathjen *et al.*, 1991, Nörenberg *et al.*, 1992). The designation tenascin-R (TN-R) has been proposed as name for the second member of the tenascin gene family and will be used for convenience in this review (Erickson, 1993b). Biochemically, TN-R glycoproteins, comparable to TN-C, assemble to multimers under native conditions, and di-, tri and tetramers have been documented (Pesheva *et al.*, 1989, Rathjen *et al.*, 1991). Increasing evidence, as discussed below, suggests that TN-R, comparable to TN-C, plays a role in mediating glial influences on neuronal and glial cell behaviour.

TN-R displays a more restricted expression than TN-C in that it seems confined to the CNS. Its major source are oligodendrocytes and its expression begins around E17 in the rat and hence is preceded by TN-C. Peak levels are reached between P8 and P15 during mouse and around E16 in chicken CNS development, in correlation with the formation of myelin loops (Rathjen *et al.*, 1991, Wintergerst *et al.*, 1993). Expression of TN-R has been assigned to oligodendrocytes both *in vivo* and *in vitro*. Recently, the glycoprotein has also been detected on axonal fiber tracts and neuron surfaces in the spinal cord and in the retinotectal projection and the TN-R message has been visualised in a minority of retinal and spinal cord neurons (Bartsch *et al.*, 1993, Fuss *et al.*, 1993, Nörenberg *et al.*, 1992, Pesheva *et al.*, 1989, Wintergerst *et al.*, 1993, Rathjen *et al.*, 1991). The significance of this co-localization is presently unclear. The regulation of TN-R expression has barely been examined. A recent study shows that astrocyte-conditioned culture supernatants, bFGF and platelet-derived growth factor stimulate, whereas DRG and spinal cord neurons reduce, TN-R expression on more differentiated O1 and O10-positive oligodendrocyte surfaces (Jung *et al.*, 1993). In juvenile zebra finch songbirds, exogeneous application of testosterone accelerates the expression of TN-R in the nucleus of the archistriatum (Burkhardt-Holm *et al.*, 1995).

FUNCTIONAL PROPERTIES OF TN-R

Comparable to TN-C, TN-R is also involved in cell-cell interactions. Mouse TN-R antibodies inhibit the binding of neurons and oligodendrocytes to oligodendrocyte surfaces in short term assays. This suggests that TN-R mediates both homotypic and heterotypic cell binding interactions. Interestingly, TN-R-mediated neuron adhesion to oligodendrocytes shifts to repulsion with increasing duration of this interaction, and, consequently, TN-R antibodies increase rather than decrease the fraction of neurons bound to oligodendrocyte surfaces during long term incubations (Morganti *et al.*, 1990). In contrast, no repulsive events are observed in the course of TN-R-dependent homophilic oligodendrocyte-oligodendrocyte interactions and oligodendrocytes grow on janusin substrates (Morganti *et al.*, 1990, Pesheva *et al.*, 1993). Duality of cell-binding and repulsion is reminiscent of analogous observations collected with TN-C (see above). This parallelism is also reflected on the structural level. Thus, chicken TN-R binds embryonic retinal neurons in cell-attachment assays and the binding site has been localized to the carboxyterminal, fibrinogen-like domain with the help of specific mAbs and rotary shadowing electron microscopy (Nörenberg *et al.*, 1992). Similarly, the carboxyterminal globular region of TN-C also harbours a cell-binding site (Figure 3.4, see above). Repulsive properties have not been reported in this system. In contrast, TN-R interferes with the binding of mouse L 929 fibroblast cells to mixed substrates of fibronectin combined with the 160 kD or 180 kD variants of mouse TN-R (Pesheva *et al.*, 1994). Furthermore, TN-R induces neuron aggregation of P6 mouse cerebellar neurons plated on TN-R/laminin mixed substrates. The repulsive effect of TN-R requires an initial attachment event apparent in short term adhesion experiments which is most pronounced after 1–2 hours. Thereafter, the fraction of cerebellar neurons detectable on TN-R substrates decreases with time in culture. In this regard, the interactions of neurons with purified TN-R adsorbed to a solid carrier evolve in two phases, comparable to the situation described for the adhesion of neurons to oligodendrocyte surfaces, as mentioned before. Consequently, these neurons do not establish monolayer cultures on homogeneous substrates of purified TN-R alone. Rather, cerebellar neuron cell bodies and their processes, and growth cones of E6 chicken retina and E7–E8 chicken DRG explants are deflected from TN-R spots applied onto laminin or polycationic poly-lysine substrata (Pesheva *et al.*, 1989, 1993, Taylor *et al.*, 1993). Again, as already discussed in the context of TN-C, these repulsive properties for neurons depend to some extent on cell lineage, because E18 rat hippocampal neurons grow and differentiate on TN-R adsorbed to nitrocellulose (Table 3.1); also, homogeneous TN-R/poly-ornithine substrates promote axon outgrowth by this class of neurons (Lochter *et al.*, 1994). Both the binding of cerebellar neurons to purified TN-R in short term adhesion assays and the deflection of growth cones from TN-R-containing areas in choice experiments is abolished by antibodies to F3. The adhesion molecule F3 is the mouse homologue of chicken F11/contactin and belongs to the Ig-superfamily of adhesion molecules. Current evidence would suggest that this GPI-linked glycoprotein is an integral element in the pathway of neuron repulsion triggered by TN-R (Gennarini *et al.*, 1989, 1991, Pesheva *et al.*, 1993). Interestingly, chicken TN-R co-purifes with F11 which is considered a neuronal receptor of the glycoprotein in that species (Brümmendorf *et al.*, 1989, Rathjen *et al.*, 1991, Brummendorf *et al.*, 1993). It is noteworthy that contactin, a homologue of F11, has also been suggested as receptor for TN-C (Ranscht and Dours, 1988, Zisch *et al.*, 1992). Which functions these glycoproteins mediate in the chicken nervous system remains to be elucidated. Summarising at this point, the available data point to interactions between the Ig-superfamily members F11/contactin/F3 and the ECM molecules TN-R and TN-C.

CONCLUSIONS AND PERSPECTIVES

In conclusion, both TN-C and TN-R are multifunctional, multimodular glycoproteins of the extracellular matrix. Several motifs have emerged from the analysis of these genes which have stimulated ongoing investigations concerning their biological functions. One salient trait resides in the topologically and temporally restricted expression of both TN-C and TN-R, e.g. in the developing nervous system (Faissner *et al.*, 1994, 1995, Faissner and Schachner, 1995). In particular, the boundary-like distribution of TN-C in some regions of the developing nervous system has been studied in detail (Steindler, 1993). In order to shed light on the potential functions of tenascin glycoproteins in neural tissues, a reductionist approach has been chosen by several laboratories. Various *in vitro* bioassays have consistently disclosed a second prominent feature of tenascin glycoproteins, namely the ambivalence between stimulatory, conducive effects on the one and inhibitory, adverse effects on the other side (Faissner, 1993, Faissner *et al.*, 1994, 1995, Faissner and Schachner, 1995). Thus, both TN-C and TN-R possess cell binding, but also repulsive sites for several cell types including neurons. Likewise, these tenascins stimulate neurite growth in some cases, but also deflect growth cones at sharp, step gradient-like boundaries, or even occasionally prevent fiber formation *in vitro*. It seems clear that the various effects of tenascin glycoproteins are caused by distinct domains. The responses to these domains presumably depend on lineage and differentiation stage of a given cell type. This might reflect the differential expression of domain-specific receptor systems and/or variations of downstream coupled intracellular effector mechanisms. The precise description of the receptors controlling neural cell responses to tenascins is still pending for most of the events detailed above, and the same holds true for 2nd-messenger pathways which are presumably activated in this context (Doherty and Walsh, 1994).

In view of the impressive range of functional effects which have been described for TN-C and TN-R in reductionist bioassays, it was not expected that the removal of the TN-C gene by homologous recombination would result in an apparently normal phenotype (Saga *et al.*, 1992). Moreover, detailed neuroanatomical studies revealed that the barrel field is preserved in the mutant, as recently assessed by monitoring its boundaries with antibodies to the chondroitin sulfate proteoglycan DSD-1-PG which is enriched in the interbarrel septae (Faissner *et al.*, 1994b, Steindler *et al.*, 1995). The reactions to the gene knock-out experiment embraced a range of attitudes comprising scepticism with respect to the role of the whole family (Erickson, 1993a) to moderate views which stressed the possibility of subtle effects. Also, the concept that vital functions such as axon guidance and the hypothesized boundaries are protected by the synergism of several functionally and/or structurally related, although not identical, molecular systems has been envisaged. A detailed and comprehensive discussion is provided in a recent review volume devoted to this issue (Tucker, 1994). The co-existence of conducive, cell-binding, growth promoting properties on the one hand and of inhibitory, repulsive qualities on the other hand has hitherto only been noted for ECM components. Recently, also laminin and merosin, ECM molecules which possess binding sites for neuronal cell bodies and growth cones (Sanes, 1989, Reichardt and Tomaselli, 1991) have been assigned anti-adhesive properties (Calof and Lander, 1991). Repulsive properties of extracellular matrix molecules might be required to regulate the intensity of cellular interactions with the pericellular environment. Release of binding forces could be helpful for cell motility and this might be the proeminent biological function of components like TN-C and TN-R. In the light of these considerations TN-C glycoproteins have also been adressed as *modulators of cell adhesion*, together with osteonectin/sparc and thrombospondin, which display counteradhesive qualities in several non-neural tissue systems (reviewed in Sage and Bornstein 1991). Modulatory recognition molecules might be important in the context of remodelling of organs and thus contribute to the response of neural tissue to injury.

ACKNOWLEDGEMENTS

The author thanks the German Research Council (DFG) for grant support (SFB 317/A2 and Fa-159/5-1,2 to AF), Dr. J. Trotter for stimulating discussion, Prof. Dr. W.B. Huttner for ongoing support, and A. Summerfield for helping prepare the artwork. Andreas Faissner is recipient of a Hermann- and Lilly-Schilling-Stiftung professorship for Neuroscience.

REFERENCES

1. Ard, M.D., Schachner, M., Rapp, J.T. and Faissner, A. (1993) Growth and degeneration of axons on astrocyte surfaces: effects on extracellular matrix and on later axonal growth. *GLIA*, 9, 248–259.

2. Atashi, J.R., Ingraham, C.A., Klinz, S.G., Matten, W.T., Schachner M. and Maness, P.F. (1992) Neural cell adhesion molecules modulate tyrosine phosphorylation of tubulin in nerve growth cone membranes. *Neuron*, **8**, 831–842.
3. Aufderheide, E., Chiquet-Ehrismann, R. and Ekblom, P. (1987) Epithelial-mesenchymal interactions in the developing kidney lead to expression of tenascin in the mesenchyme. *J. Cell Biol.*, **105**, 599–608.
4. Aufderheide, E. and Ekblom, P. (1988) Tenascin during gut development : appearence in the mesenchyme, shift in molecular forms, and dependence on epithelial-mesenchymal interactions. *J. Cell Biol.*, **107**, 2341–2349.
5. Aukhil, I., Joshi, P., Yan, Y. and Erickson, H.P. (1993) Cell- and heparin-binding domains of the hexabrachion arm identified by tenascin expression proteins. *J. Biol. Chem.*, **268**, 2542–2553.
6. Bachmann, M., Conscience, J.F., Probstmeier, R., Carbonetto, S. and Schachner, M. (1995) Recognition molecules myelin-associated glycoprotein and tenascin-C inhibit integrin-mediated adhesion of neural cells to collagen. *J. Neurosci. Res.*, **40**, 458–470.
7. Baier, H. and Bonhoeffer, F. (1992) Axon guidance by gradients of a target-derived component. *Science*, **255**, 472–475.
8. Bandtlow, C., Zachleder, T. and Schwab, M. E. (1990) Oligodendrocytes arrest neurite growth by contact inhibition. *J. Neurosci.*, **10**, 3837–3848.
9. Bandtlow, C.E., Schmidt, M.F., Hassinger, T.D., Schwab, M.E. and Kater, S.B. (1993) Role of intracellular calcium in NI-35-evoked collapse of neuronal growth cones. *Science*, **259**, 80–83.
10. Barnea, G., Grumet, M., Milev, P., Silvennoinen, O., Levy, J.B., Sap, J. and Schlessinger, J. (1994) Receptor tyrosine phosphatase β is expressed in the form of proteoglycan and binds to the extracellular matrix protein tenascin. *J. Biol. Chem.*, **269**, 14349–14352.
11. Bartsch, S., Bartsch, U., Dörries, U., Faissner, A., Weller, A., Ekblom, P. and Schachner, M. (1992) Expression of tenascin in the developing and adult cerebellar cortex. *J. Neurosci.*, **12**, 736–749.
12. Bartsch, S., Husmann, K., Schachner, M. and Bartsch, U. (1995) The extracellular matrix molecule tenascin: expression in the developing chick retinotectal system and substrate properties for retinal ganglion cell neurites *in vitro*. *Europ. J. Neurosci.*, **7**, 907–916.
13. Bartsch, U., Bartsch, S., Dörries, U. and Schachner, M. (1992) Immunohistological localization of tenascin in the developing and lesioned adult mouse optic nerve. *Europ. J. Neurosci.*, **4**, 338–352.
14. Bartsch, U., Pesheva, P., Raff, M. and Schachner, M. (1993) Expression of janusin (J1-160/180) in the retina and optic nerve of the developing and adult mouse. *GLIA*, **9**, 57–69
15. Bartsch, U., Faissner, A., Trotter, J., Dörries, U., Bartsch, S., Mohajeri, H. and Schachner, M. (1994) Tenascin demarcates the boundary between myelinated and non-myelinated retinal ganglion cell axons of the adult mouse. *J. Neurosci.*, **14**, 4756–4768.
16. Bastmeyer, M., Beckmann, M., Schwab, M.E. and Stuermer, C.A.O. (1991) Growth of regenerating goldfish axons is inhibited by rat oligodendrocytes and CNS myelin but not by goldfish optic nerve tract oligodendrocyte-like cells and fish CNS myelin. *J. Neurosci.*, **11**, 626–640.

17. Bastmeyer, M. and Stuermer, C.A.O. (1993) Behavior of fish retinal growth cones encountering chick caudal tectal membranes: a time-lapse study on growth cone collapse. *J. Neurobiol.*, **24**, 37–50.
18. Bixby, J.L. and Harris, W.A. (1991) Molecular mechanisms of axon growth and guidance. *Annu. Rev. Cell Biol.* **7**, 117–159.
19. Bourdon, M.A., Wikstrand, C.J., Furthmayr, H., Matthews, T.J. and Bigner, D.D. (1983) Human glioma-mesenchymal extracellular matrix antigen defined by monoclonal antibody. *Cancer Res.*, **43**, 2796–2805.
20. Bourdon, M.A., Matthews, T.H., Pizzo, S.V. and Bigner, D.D. (1985) Immunochemical and biochemical characterization of a glioma-associated extracellular matrix glycoprotein. *J. Cell. Biochem.*, **28**, 183–195.
21. Bourdon, M.A. and Ruoslahti, E. (1989) Tenascin mediates cell attachment through an RGD-dependent receptor. *J. Cell Biol.*, **108**, 1149–1155.
22. Brodkey, J.A., Laywell, E.D., O'Brien, T.F., Faissner, A., Stefansson, K., Dörries, U., Schachner, M. and Steindler, D.A. (1995) Focal brain injury and upregulation of a developmentally-regulated extracellular matrix protein. *J. Neurosurgery*, **82**, 106–112.
23. Bronner-Fraser, M. (1988) Distribution and function of tenascin during cranial neural crest development in the chick. *J. Neurosci. Res.*, **21**, 135–147.
24. Brümmendorf, T., Wolff, J.M., Frank, R. and Rathjen, F.G. (1989) Neural cell recognition molecule F11: homology with fibronectin type III and immunoglobulin type C domains. *Neuron*, **2**, 1351–1361.
25. Brümmendorf, T., Hubert, M., Treubert, U., Leuschner, R., Tarnok, A. and Rathjen, F.G. (1993) The axonal recognition molecule F11 is a multifunctional protein: specific domains mediate interactions with Ng-CAM and restrictin. *Neuron*, **10**, 711–727.
26. Brümmendorf, T. and Rathjen, F.G. (1993) Axonal glycoproteins with immunoglobulin- and fibronectin type III-related domains in vertebrates: structural features, binding activities, and signal transduction. *J. Neurochem.*, **61**, 1207–1219.
27. Burkhardt-Holm, P., Kafitz, K.W., Güttinger, H.R. and Schachner, M. (1995) Exogenous testosterone alters expression pattern of cell recognition molecules in brains of juvenile zebra finches. *Neuroreport*, **6**, 433–436.
28. Calof, A.L. and Lander, A.D. (1991) Relationship between neuronal migration and cell-substratum adhesion: laminin and merosin promote olfactory neuronal migration but are anti-adhesive. *J. Cell Biol.*, **115**, 779–794.
29. Caroni, P. and Schwab, M.E. (1988a) Antibody against myelin-associated inhibitor of neurite growth neutralizes nonpermissive substrate properties of CNS white matter. *Neuron*, **1**, 85–96.
30. Caroni, P. and Schwab, M.E. (1988b) Two membrane protein fractions from rat central myelin with inhibitory properties for neurite growth and fibroblast spreading. *J. Cell Biol.*, **106**, 1281–1288.
31. Caubit, X., Riou, J.F., Coulon, J., Arsanto, J.P., Benraiss, A., Boucaut, J.C. and Thouveny, Y. (1994) Tenascin expression in developing, adult and regenerating caudal spinal cord in the urodele amphibians. *Int. J. Dev. Biol.*, **38**, 661–672.
32. Celio, M.R. and Blümcke, I. (1994) Perineuronal nets — a specialized form of extracellular matrix in the adult nervous system. *Brain Res. Revs.*, **19**, 128–145.

33. Chiquet, M., Vrucinic-Filipi, N., Schenk, S., Beck, K. and Chiquet-Ehrismann, R. (1991) Isolation of chick tenascin variants and fragments. *Eur. J. Biochem.*, **199**, 379–388.
34. Chiquet, M. and Fambrough, D.M. (1984a) Chick myotendinous antigen. I. A monoclonal antibody as a marker for tendon and muscle morphogenesis. *J. Cell Biol.*, **98**, 1926–1936.
35. Chiquet, M. and Fambrough, D.M. (1984b) Chick myotendinous antigen. II. A novel extracellular glycoprotein complex consisting of large disulfide-linked subunits. *J. Cell Biol.*, **98**, 1937–1946.
36. Chiquet-Ehrismann, R. (1991) Anti-adhesive molecules of the extracellular matrix. *Curr. Opin. Cell Biol.*, **3**, 800 – 804.
37. Chiquet-Ehrismann, R., Macke, E.J., Pearson, C.A. and Sakakura, T. (1986) Tenascin: an extracellular matrix protein involved in tissue interactions during fetal development and oncogenesis. *Cell*, **47**, 131–139.
38. Chiquet-Ehrismann, R., Kalla, P., Pearson, C.A., Beck, K. and Chiquet, M. (1988) Tenascin interfers with fibronectin action. *Cell*, **53**, 383 –390.
39. Chiquet-Ehrismann, R., Tannheimer, M., Koch, M., Brunner, A., Spring, J , Martin, D., Baumgartner, S. and Chiquet, M. (1994) Tenascin-C expression by fibroblasts is elevated in stressed collagen gels. *J. Cell Biol.*, **127**, 2093 –2101.
40. Chou, D.K.H., Ilyas, A.A., Evans, J.E., Costello, C., Quarles, R.H. and Jungalwala, F.B. (1986) Structure of sulfated glucuronyl glycolipids in the nervous system reacting with HNK-1 antibody and some IgM paraproteins in neuropathy. *J. Biol. Chem.*, **25**, 11717–11725.
41. Chung, C.Y. and Erickson, H.P. (1994) Cell surface annexin II is a high affinity receptor for the alternatively spliced segment of tenascin-C. *J. Cell Biol.*, **126**: 539–548.
42. Chuong, C.-M., Crossin, K.L. and Edelman, G.M. (1987) Sequential expression and differential function of multiple adhesion molecules during the formation of cerebellar cortical layers. *J. Cell Biol.*, **104**, 331–342.
43. Cooper, N.G.F. and Steindler, D.A. (1986a) Lectins demarcate the barrel subfield in the somatosensory cortex of the early postnatal mouse. *J. Comp. Neurol.*, **249**, 157–168.
44. Cooper, N.G.F. and Steindler, D.A. (1986b) Monoclonal antibody to glial fibrillary acidic protein reveals a parcellation of individual barrels in the early postnatal mouse somatosensory cortex. *Brain Res.*, **380**, 341–348.
45. Cooper, N.G.F. and Steindler, D.A. (1989) Critical period-dependent alterations of the transient body image in the rodent cerebral cortex. *Brain Res.*, **489**, 167–176.
46. Copertino, D.W., Jenkinson, S., Jones, F.S. and Edelman, G.M. (1995) Structural and functional similarities between the promoters for mouse tenascin and chicken cytotactin. *Proc. Natl. Acad. Sci. USA*, **92**, 2131–2135.
47. Cox, E.C., Müller, B. and Bonhoeffer, F. (1990) Axonal guidance in the chick visual system: posterior tectal membranes induce collapse of growth cones from the temporal retina. *Neuron*, **2**, 31–37.
48. Crossin, K.L., Hoffman, S., Grumet, M., Thiery, J.-P. and Edelman, G.M. (1986) Site-restricted expression of cytotactin during development of the chicken embryo. *J. Cell Biol.*, **102**, 1917–1930.

49. Crossin, K.L., Hoffman, S., Tan, S.-S. and Edelman, G.M. (1989) Cytotactin and its proteoglycan ligand mark structural and functional boundaries in somatosensory cortex of the early postnatal mouse. *Dev. Biol.*, **136**, 381–392.
50. Crossin, K.L., Prieto, A., Hoffman, S., Jones, F.S. and Friedlander, D. (1990) Expression of adhesion molecules and the establishment of boundaries during embyonic and neural development. *Exp. Neurol.*, **109**, 6–18.
51. Crossin, K.L. (1991) Cytotactin binding: inhibition of stimulated proliferation and intracellular alkalinization in fibroblasts. *Proc. Natl. Acad. Sci. USA*, **88**, 11403–11407.
52. Crossin, K.L. and Edelman, G.M. (1992) Specific binding of cytotactin to sulfated glycolipids. *J. Neurosci. Res.*, **33**, 631–638.
53. Daniloff, J.K., Crossin, K.L., Pincon-Raymond, M., Murawsky, M., Rieger, F. and Edelman, G.M. (1989) Expression of cytotactin in the normal and regenerating neuromuscular system. *J. Cell Biol.*, **108**, 625–635.
54. Davies, J.A., Cook, G.M.W., Stern, C.D. and Keynes, R.J. (1990) Isolation from chick somites of a glycoprotein fraction that causes collapse of dorsal root ganglion growth cones. *Neuron*, **2**, 11–20.
55. Dodd, J. and Schuchardt, A. (1995) Axon guidance: a compelling case for repelling growth cones. *Cell*, **81**, 471–474.
56. Doherty, P. and Walsh, F.S. (1994) Signal transduction events underlying neurite outgrowth stimulated by cell adhesion molecules. *Curr. Opin. Neurobiol.*, **4**, 49–55.
57. Edelman, G.M. and Jones, F.S. (1992) Cytotactin: a morphoregulatory molecule and a target for regulation by homeobox gene products. *Trends Biochem.*, **17**, 228–232.
58. Ekblom, P. and Aufderheide E. (1989) Stimulation of tenascin expression in mesenchyme by epithelial-mesenchymal interactions. *Int. J. dev. Biol.*, **33**, 71–79.
59. Ekblom, M., Fässler, R., Tomasini-Johansson, B., Nilsson, K. and Ekblom, P. (1993) Downregulation of tenascin expression by glucocorticoids in bone marrow stromal cells and in fibroblasts. *J. Cell Biol.*, **123**, 1037–1045.
60. Epperlein, H.H., Halfter, W. and Tucker, R.P. (1988) The distribution of fibronectin and tenascin along migratory pathways of the neural crest in the trunk of amphibian embryos. *Development*, **103**, 743–756.
61. Erickson, H.P. and Inglesias, J.L. (1984) A six-armed oligomer isolated from cell surface fibronectin preparations. *Nature*, **311**, 267–269.
62. Erickson, H.P. and Taylor, H C. (1987) Hexabrachion proteins in embryonic chicken tissues and human tumors. *J. Cell Biol.*, **105**, 1387–1394.
63. Erickson, H.P. and Bourdon, M.A. (1989) Tenascin: an extracellular matrix protein prominent in specialized embryonic tissues and tumors. *Annu. Rev. Cell Biol.*, **5**, 71–92 (1989)
64. Erickson, H.P. (1993a) Gene knockouts of c-src, transforming growth factor β-1, and tenascin suggest superfluous, nonfunctional expression of proteins. *J. Cell Biol.*, **120**, 1079–1081.
65. Erickson, H.P. (1993b) Tenascin-C, tenascin-R and tenascin-X: a family of talented proteins in search of functions. *Curr. Opin. Cell Biol.*, **5**, 869–876.

66. Faissner, A., Kruse, J., Chiquet-Ehrismann, R. and Mackie, E.J. (1988) The high-molecular-weight-J1-glycoproteins are immunochemically related to Tenascin. *Differentiation*, **37**, 104–114.
67. Faissner, A., Kruse, J., Kühn, K. and Schachner, M. (1990) Binding of the J1-adhesion molecules to extracellular matrix constituents. *J. Neurochem.*, **54**, 1004–1015.
68. Faissner, A. and Kruse, J. (1990) J1/tenascin is a repulsive substrate for central nervous system neurons. *Neuron*, **5**, 627–637.
69. Faissner, A. (1993) Tenascin glycoproteins in neural pattern formation - facets of a complex picture. *Persp. Dev. Neurobiol.*, **1**, 155–164.
70. Faissner, A., Scholze, A. and Götz, B. (1994a) Tenascin glycoproteins in developing neural tissues – only decoration? *Persp. Dev. Neurobiol.*, **2**, 53–66.
71. Faissner, A., Clement, A., Lochter, A., Streit, A., Mandl, C. and Schachner, M. (1994b) Isolation of a neural chondroitin sulfate proteoglycan with neurite outgrowth promoting properties. *J. Cell Biol.*, **126**, 783–799.
72. Faissner, A., Götz, B., Joester, A. and Scholze, A. (1995) The tenascin gene family – versatile glycoproteins implicated in neural pattern formation and regeneration. *Semin. Dev. Biol.*, **6**, 139–148.
73. Faissner, A. and Schachner, M. (1995) Tenascin and janusin – glial recognition molecules involved in neural development and regeneration. In: *Neuroglial Cells*, H. Kettenmann and B.R. Ransom eds., Oxford University Press, Oxford, N. York, in press.
74. Faissner, A. and Steindler, D.A. (1995) Boundaries and inhibitory molecules in developing neural tissues. *GLIA*, **13**, 233–254.
75. Fawcett, J., Rokos, J. and Bakst, I. (1989) Oligodendrocytes repel axons and cause growth cone collapse. *J. Cell Sci.*, **92**, 93–100.
76. Fischer, E.H., Charbonneau, H. and Tonks, N.K. (1991) Protein tyrosine phosphatases: a diverse family of intracellular and transmembrane enzymes. *Science*, **253**, 401–406.
77. Fischer, D., Chiquet-Ehrismann, R., Bernasconi, C. and Chiquet, M. (1995) A single heparin binding region within the fibrinogen-like domain is functional in chicke tenascin-C. *J. Biol. Chem.*, **270**, 3378–3384.
78. Friedlander, D.R., Hoffman, S. and Edelman, G.M. (1988) Functional mapping of cytotactin: proteolytic fragments active in cell substrate adhesion. *J. Cell Biol.*, **107**, 2329–2340.
79. Friedlander, D.R., Milev, P., Karthikeyan, L., Margolis, R.K., Margolis, R.U. and Grumet, M. (1994) The neuronal chondroitin sulfate proteoglycan neurocan binds to the neural cell adhesion molecules Ng-CAM/L1/NILE and N-CAM, and inhibits neuronal adhesion and neurite outgrowth. *J. Cell Biol.*, **125**, 669–680.
80. Fruttiger, M., Schachner, M. and Martini, R. (1995) Tenascin-C expression during Wallerian degeneration in C57BL/Wld(S) mice: possible implications for axonal regeneration. *J. Neurocytol.*, **24**, 1–14.
81. Fuss, B., Wintergerst, E.S., Bartsch, U. and Schachner, M. (1993) Molecular characterisation and in-situ mRNA localisation of the neural recognition molecule J1-160/180: a modular structure similar to tenascin. *J. Cell Biol.*, **120**, 1237–1249.
82. Gatchalian, C.L., Schachner, M. and Sanes, J. (1989) Fibroblasts that proliferate near denervated synaptic sites in skeletal muscle synthesize the adhesive

molecules tenascin(J1), N-CAM, fibronectin, and a heparan sulfate proteoglycan. *J. Cell Biol.*, **108**, 1873–1890.
83. Gates, M.A., Thomas, B.L., Howard, E.M., Laywell, E.D., Sajin, B., Faissner, A., Götz, B., Silver, J. and Steindler, D.A. (1995) Cell and molecular analysis of the developing and adult mouse subventricular zone of the cerebral hemispheres. *J. Comp. Neurol.*, in press.
84. Geiger, B. and Ayalon, O. (1992) Cadherins. *Annu. Rev. Cell Biol.*, **8**, 307–332.
85. Gennarini, G., Cibelli, G., Rougon, G., Mattei, M.-G. and Goridis, C. (1989) The mouse neuronal cell surface glycoprotein F3: a phosphatidyl inositol-anchored member of the immunoglobulin superfamily related to chicken contactin. *J. Cell Biol.*, **109**, 775–788.
86. Gennarini, G., Durbec, P., Boned, A., Rougon, G. and Goridis, C. (1991) Transfected F3/F11 neuronal cell surface protein mediates intercellular adhesion and promotes neurite outgrowth. *Neuron*, **6**, 596–606.
87. Gherzi, R., Carnemolla, B., Siri, A., Ponassi, M., Balza, E. and Zardi, L. (1995) Human tenascin gene — structure of the 5′-region, identification, and characterization of the transcription regulatory sequences. *J. Biol. Chem.*, **270**, 3429–3434.
88. Godfraind, C., Schachner, M. and Goffinet, A.M. (1988) Immunohistological localization of cell adhesion molecules L1, J1, N-CAM and their common carbohydrate L2 in the embryonic cortex of normal and reeler mice. *Dev. Brain Res.*, **42**, 99–111.
89. Gonzalez, M. de L., Malemud, C.J. and Silver, J. (1993) Role of astroglial extracellular matrix in the formation of rat olfactory bulb glomeruli. *Exp. Neurol.*, **123**, 91–105.
90. Gonzalez, M. de L. and Silver, J. (1993) Axon-glia interactions regulate ECM patterning in the postnatal rat olfactory bulb. *J. Neurosci.*, **14**, 6121–6131.
91. Grierson, J.P., Petroski, R.E., Ling, D.S.F. and Geller, H. (1990) Astrocyte topography and tenascin/cytotactin expression: correlation with the ability to support neuritic outgrowth. *Dev. Brain Res.*, **55**, 11–19.
92. Grumet, M., Hoffman, S., Crossin, K.L. and Edelman, G.M. (1985) Cytotactin, an extracellular matrix protein of neural and non-neural tissues that mediates glia-neuron interaction. *Proc. Natl. Acad. Sci. USA*, **82**, 8075–8079.
93. Grumet, M., Milev, P., Sakurai, T., Karthikeyan, L., Bourdon, M., Margolis, R.K. and Margolis, R. (1994) Interactions with tenascin and differential effects on cell adhesion of neurocan and phosphacan, two major chondroitin sulfate proteoglycans of nervous tissue. *J. Biol. Chem.*, **269**, 12142–12146.
94. Gulcher, J.R., Marton, L.S. and Stefansson, K. (1986) Two large glycosylated polypeptides found in myelinating oligodendrocytes but not in myelin. *Proc. Natl. Acad. Sci. USA*, **83**, 2118–2122.
95. Gulcher, J.R., Nies, D.E., Marton, L.S. and Stefansson, K. (1989) An alternatively spliced region of the human hexabrachion contains a repeat of potential N-glycosylation sites. *Proc. Natl. Acad. Sci. USA*, **86**, 1588–1592.
96. Halfter, W., Chiquet-Ehrismann, R. and Tucker, R.P. (1989) The effect of tenascin and embryonic basal lamina on the behavior and morphology of neural crest cells *in vitro*. *Dev. Biol.*, **132**, 14–25.
97. Hassel, J.R., Kimura, J.H. and Hascall, V.C. (1986) Proteoglycan core protein families. *Annu. Rev. Biochem.*, **55**, 539–567.

98. Herndon, M.E. and Lander, A.D. (1990) A diverse set of developmentally regulated proteoglycans is expressed in the rat central nervous system. *Neuron*, **4**, 949–961.
99. Higuchi, M., Ohnishi, T., Arita, N., Hiraga, S. and Hayakawa, T. (1993) Expression of tenascin in human gliomas: its relation to histological malignancy, tumor dedifferentiation and angiogenesis. *Acta Neuropatol.*, **85**, 481–487.
100. Hoffman, S. and Edelman, G.M. (1987) A proteoglycan with HNK-1 antigenic determinants is a neuron-associated ligand for cytotactin. *Proc. Natl. Acad. Sci. USA*, **84**, 2523–2527.
101. Hoffman, S., Crossin, K.L. and Edelman, G.M. (1988) Molecular forms, binding functions, and developmental expression patterns of cytotactin and cytotactin-binding proteoglycan, an interactive pair of extracellular matrix molecules. *J. Cell. Biol.*, **106**, 519–532.
102. Husmann, K., Faissner, A. and Schachner, M (1992) Tenascin promotes cerebellar granule cell migration and neurite outgrowth by different domains in the fibronectin type III repeats. *J. Cell Biol.*, **116**, 1475–1486.
103. Hynes, R.O. and Lander, A.D. (1992) Contact and adhesive specificities in the associations, migrations, and targeting of cells and axons. *Cell*, **68**, 303–322.
104. Hynes, R.O. (1992) Integrins: versatility, modulation, and signaling in cell adhesion. *Cell*, **69**, 11–25.
105. Igarashi, M., Strittmater, S.M., Vartanian, T. and Fishman, M.C. (1993) Mediation by G proteins of signals that cause collapse of growth cones. *Science*, **259**, 77–79.
106. Irintchev, A., Salvini, T.F., Faissner, A. and Wernig, A. (1993) Differential expression of tenascin after denervation, damage or paralysis of mouse soleus muscle. *J. Neurocytol.*, **22**, 955–965.
107. Jacobson, M. (1991) *Developmental Neurobiology*, 3rd Edition, Plenum Press, New York and London, pp. 1–776.
108. Jessel, T.M. (1988) Adhesion molecules and the hierarchy of neural development. *Neuron*, **1**, 3–13.
109. Jhaveri, S., Erzurumlu, R.S. and Crossin, K. (1991) Barrel construction in rodent neocortex: role of thalamic afferents versus extracellular matrix molecules. *Proc. Natl. Acad. Sci. USA*, **88**, 4489–4493.
110. Jones, F.S., Hoffman, S., Cunningham, B.A. and Edelman, G.M. (1989) A detailed structural model of cytotactin: protein homologies, alternative RNA splicing, and binding regions. *Proc. Natl. Acad. Sci USA*, **86**, 1905–1909.
111. Jones, F.S., Crossin, K.L., Cunningham, B.A. and Edelman, G.M. (1990) Identification and characterization of the promoter for the cytotactin gene. *Proc. Natl. Acad. Sci. USA*, **87**, 6497–6501.
112. Jones, F.S., Chalepakis, G., Gruss, P. and Edelman, G.M. (1992) Activation of the cytotactin promoter by the homeobox-containing gene Evx-1. *Proc. Natl. Acad. Sci. USA*, **89**, 2091–2095.
113. Joshi, P., Chung, C.Y., Aukhil, I. and Erickson, H.P. (1993) Endothelial cells adhere to the RGD domain and the fibrinogen-like terminal knob of tenascin. *J. Cell Sci.*, **106**, 389–400.
114. Jung, M., P. Pesheva, M. Schachner and J. Trotter. (1993) Astrocytes and neurons regulate the expression of the neural recognition molecule janusin by cultured oligoden-drocytes. *GLIA*, **9**, 163–175.

115. Kapfhammer, J.P. and Raper, J.A. (1987a) Collapse of growth cone structure on contact with specific neurites in culture. *J. Neurosci.*, **7**, 201–212.
116. Kapfhammer, J.P. and Raper, J.A. (1987b) Interactions between growth cones and neurites growing from different neural tissues in culture. *J. Neurosci.*, **7**, 1595–1600.
117. Kaplony, A., Zimmermann, D.R., Fischer, R.W., Imhof, B.A., Odermatt, B.F., Winterhalter, K.H. and Vaughan, L. (1991) Tenascin Mr 220 000 isoform expression correlates with corneal cell migration. *Development*, **112**, 605–614.
118. Kawano, H., Ohyama, K., Kawamura, K. and Nagatsu, I (1995) Migration of dopaminergic neurons in the embryonic mesencephalon of mice. *Dev. Brain Res.*, **86**, 101–113.
119. Keilhauer, G., Faissner, A. and Schachner, M. (1985) Differential inhibition of neuron-neuron, neuron-astrocyte and astrocyte-astrocyte adhesion by L1, L2 and N-CAM antibodies. *Nature*, **316**, 728–730.
120. Keynes, R.J. and Stern, C.D. (1984) Segmentation in the vertebrate nervous system. *Nature*, **310**, 786–789.
121. Keynes, R. and Lumsden, A. (1990) Segmentation and the origin of regional diversity in the vertebrate central nervous system. *Neuron*, **2**, 1–9.
122. Keynes, R., and Cook, G.M.W. (1992) Repellent cues in axon guidance. *Curr. Opinions Neurobiol.*, **2**, 55–59.
123. Kiernan, B.W. and ffrench-Constant (1993) Oligodendrocyte precursor (O-2A progenitor cell) migration; a model system for the study of cell migration in the developing central nervous system. *Development Supplement*, 219–225.
124. Kolodkin, A.L., Matthes, D.J., O'Connor, T.P., Patel, N.H., Admon, A., Bentley, D. and Goodman, C.S. (1992) Fasciclin IV: sequence, expression, and function during growth cone guidance in the grasshopper embryo. *Neuron*, **9**, 831–845.
125. Kolodkin, A.L., Matthes, D.J. and Goodman, C.S. (1993) The semaphorin genes encode a family of transmembrane and secreted growth cone guidance molecules. *Cell*, **75**, 1389–1399.
126. Krull, C.E., Morton, D.B., Faissner, A., Schachner, M. and Tolbert, L. (1994a) Spatiotemporal pattern of expression of tenascin-like molecules in a developing insect olfactory system. *J. Neurobiol.*, **5**, 515–534.
127. Krull, C.E., Oland, L.A., Faissner, A., Schachner, M. and Tolbert, L.E. (1994b) In vitro analyses of neurite outgrowth indicate a potential role for tenascin-like molecules in the development of insect olfactory glomeruli. *J. Neurobiol.*, **25**, 989–1004.
128. Kruse, J., Mailhammer, R., Wernecke, H., Faissner, A., Sommer, I., Goridis, C. and Schachner, M. (1984) Neural cell adhesion molecule and myelin associated glycoprotein share a common, developmentally early carbohydrate moiety recognized by monoclonal antibodies L2 and HNK-1. *Nature*, **311**, 153–155.
129. Kruse, J., Keilhauer, G., Faissner, A., Timpl, R. and Schachner, M. (1985) The J1 glycoprotein – a novel nervous system cell adhesion molecule of the L2/HNK-1 family. *Nature*, **316**, 146–148.
130. Lander, A.D. (1993) Proteoglycans in the nervous system. *Curr. Opin. Neurobiol.*, **3**, 716–723.
131. Langenfeld-Oster, B., Faissner, A., Irintchev, A. and Wernig, A. (1994) Polyclonal antibodies against N-CAM and tenascin delay endplate reinnervation. *J. Neurocytol.*, **23**, 591–604.

132. Laywell, E.D. and Steindler, D.A. (1991) Boundaries and wounds, glia and glycoconjugates. *Ann. N. Y. Acad. Sci.*, **633**, 122–144.
133. Laywell, E., Dörries, U., Bartsch, U., Faissner, A., Schachner, M. and Steindler, D.A. (1992) Enhanced expression of the developmentally regulated extracellular matrix molecule tenascin following adult brain injury. *Proc. Natl. Acad. Sci. USA*, **89**, 2634–2638.
134. Letourneau, P.C., Condic, M.L. and Snow, D.M. (1994) Interactions of developing neurons with the extracellular matrix. *J. Neurosci.*, **14**, 915–928.
135. Lightner, V.A. and Erickson, H.P. (1990) Binding of hexabrachion (tenascin) to the extracellular matrix and substratum and its effect on cell adhesion. *J. Cell Science*, **95**, 263–277.
136. Lochter, A., Vaughan, L., Kaplony, A., Prochiantz, A., Schachner, M. and Faissner, A. (1991) J1/tenascin in substrate-bound and soluble form displays contrary effects on neurite outgrowth. *J. Cell Biol.*, **113**, 1159–1171.
137. Lochter, A. and Schachner, M. (1993) Tenascin and extracellular matrix proteins – from promotion to polarization of neurite growth in vitro. *J. Neurosci.*, **13**, 3986–4000.
138. Lochter, A., Taylor, J., Fuss, B. and Schachner, M. (1994) The extracellular matrix molecule janusin regulates neuronal morphology in a substrate- and culture time-dependent manner. *Eur. J. Neurosci.*, **6**, 597–606.
139. Lotz, M.M., Burdsal, C.A., Erickson, H.P. and McClay, D.R. (1989) Cell adhesion to fibronectin and tenascin: Quantitative measurements of initial binding and subsequent strengthening response. *J. Cell Biol.*, **109**, 1795–1805.
140. Luo, Y., Raible, D. and Raper, J.A. (1993) Collapsin: a protein that induces the collapse and paralysis of neuronal growth cones. *Cell*, **75**, 217–227.
141. Luo, Y. and Raper, J.A. (1994) Inhibitory factors controlling growth cone motility and guidance. *Curr. Opin. Neurobiol.*, **4**, 648–654.
142. Mackie, E.J., Tucker, R.P., Halfter, W., Chiquet-Ehrismann, R. and Epperlein, H.H. (1988) The distribution of tenascin coincides with pathways of neural crest migration. *Development*, **102**, 237–250.
143. Martini, R., Schachner, M. and Faissner, A. (1990) Enhanced expression of the extracellular matrix molecule J1/tenascin in the regenerating adult mouse sciatic nerve. *J. Neurocytol.*, **19**, 601–616.
144. Martini, R., and Schachner, M. (1991) Complex Expression pattern of tenascin during innvervation of the posterior limb buds of the developing chicken. *J. Neurosci. Res.*, **28**, 261–279.
145. Martini, R. (1994) Expression and functional roles of neural cell surface molecules and extracellular matrix components during development and regeneration of peripheral nerves. *J. Neurocytol.*, **23**, 1–28.
146. Marton, L.S., Gulcher, J.R. and Stefansson, K. (1989) Binding of hexabrachions to heparin and DNA. *J. Biol. Chem.*, **264**, 13145–13149.
147. Mason, C.A. (1993) Guest Editor of the Special Issue *The midline as a boundary for central nervous system development. Perspec. Dev. Neurobiol.*, **1**, 179–268.
148. Masuda-Nakagawa, L.M. and Wiedemann, C. (1992) The role of matrix molecules in regeneration of Leech CNS. *J. Neurobiol.*, **23**, 551–567.
149. Maurel, P., Rauch, U., Flad, M., Margolis, R.K. and Margolis, R.U. (1994) Phosphacan, a chondroitin sulfate proteoglycan of brain that interacts with

neurons and neural cell adhesion molecules, is an extracellular variant of a receptor-type protein tyrosine phosphatase. *Proc. Natl. Acad. Sci. USA*, **91**, 2512–2516.

150. McKeon, R.J., Schreiber, R.C., Rudge, J.S. and Silver, J. (1991) Reduction of neurite outgrowth in a model of glial scarring following CNS injury is correlated with the expression of inhibitory molecules on reactive astrocytes. *J. Neurosci.*, **11**, 3398–3411.
151. Mège, R.M., Nicolet, M., Pincon-Raymond, M., Murawsky, M. and Rieger, F. (1992) Cytotactin is involved in synaptogenesis during regeneration of the frog neuromuscular system. *Dev. Biol.*, **149**, 381–394.
152. Meiners, S., Marone, M., Rittenhouse J.L. and Geller, H.M. (1993) Regulation of astrocytic tenascin by basic fibroblast growth factor. *Dev. Biol.*, **160**, 480–493.
153. Miragall, F., Kadmon, G., Faissner, A., Antonicek, H. and Schachner, M. (1990) Retention of J1/tenascin and the polysialylated form of the neural cell adhesion molecule (N-CAM) in the adult olfactory bulb. *J. Neurocytol.*, **19**, 899–914.
154. Mitrovic, N., Dörries, U. and Schachner, M. (1994) Expression of the extracellular matrix glycoprotein tenascin in the somatosensory cortex of the mouse during postnatal development: an immunocytochemical and in situ hybridization analysis. *J. Neurocytol.*, **23**, 364–378.
155. Morganti, C.M., Taylor, J., Pesheva, P. and Schachner, M. (1990) Oligodendrocyte-derived J1-160/180 extracellular matrix glycoproteins are adhesive or repulsive depending on the partner cell type and time of interaction. *Exp. Neurol.*, **109**, 98–110.
156. Murphy-Ullrich, J.E., Lightner, V.A., Aukhil, I., Yan, Y.Z., Erickson, H.P. and Höök, M. (1991) Focal adhesion integrity is downregulated by the alternatively spliced domain of human tenascin. *J. Cell Biol.*, **115**, 1127–1136.
157. Nörenberg, U., Wille, H., Wolff, J.M., Frank, R. and Rathjen, F.G. (1992) The chicken neural extracellular matrix molecule restrictin: similarity with EGF-, fibronectin type III, and fibrinogen-like motifs. *Neuron*, **8**, 849–863.
158. O'Brien, T.F., Faissner, A., Schachner, M. and Steindler, D.A. (1992) Afferent-boundary interactions in the developing neostriatal mosaic. *Dev. Brain Res.*, **65**, 259–267.
159. Oohira, A., Matsui, F. and Katho-Semba, R. (1991) Inhibitory effects of brain chondroitin sulfate proteoglycans on neurite outgrowth from PC12D cells. *J. Neurosci.*, **11**, 822–827.
160. Patterson, P.H. (1988) On the importance of being inhibited, or saying no to growth cones. *Neuron*, **1**, 263–267.
161 Pearson, C.A., Pearson, D., Shibahara, S., Hofsteenge, J., and Chiquet-Ehrismann, R. (1988) Tenascin: cDNA cloning and induction by TGF-β. *EMBO J.* **7**: 2677–2981.
162. Perez, R.G. and Halfter, W. (1993) Tenascin in the developing chick visual system: distribution and potential role as a modulator of retinal axon growth. *Dev. Biol.*, **156**, 278–292.
163. Pesheva, P., Spiess, E. and Schachner, M. (1989) J1-160 and J1-180 are oligodendrocyte-secreted nonpermissive substrates for cell adhesion. *J. Cell Biol.*, **109**, 1765–1778.
164. Pesheva, P., Gennarini, G., Goridis, C. and Schachner, M. (1993) The FIII/F11 cell adhesion molecule mediates the repulsion of neurons by the extracellular

matrix glycoprotein J1–160/180. *Neuron*, **10**, 69–82.

165. Pesheva, P., Probstmeier, R., Skubitz, A.P.N., McCarthy, J.B., Furcht, L.T. and Schachner, M. (1994) Tenascin-R (J1 160/180) inhibits fibronectin-mediated cell adhesion - functional relatedness to tenascin-C. *J. Cell Sci.*, **107**, 2323–2333.
166. Pindzolla, R.R., Doller, C. and Silver, J. (1993) Putative inhibitory extracellular matrix molecules at the dorsal root entry zone of the spinal cord during development and after root and sciatic nerve lesions. *Dev. Biol.*, **156**, 34–48.
167. Prieto, A.L., Jones, F.S., Cunningham, B.A., Crossin, K.L. and Edelman, G.M. (1990) Localization during development of alternatively spliced forms of cytotactin mRNA by in situ hybridization. *J. Cell Biol.*, **111**, 685–698.
168. Prieto, A.L., Andersson-Fisone, C. and Crossin, K.L. (1992) Characterisation of multiple adhesive and counteradhesive domains in the extracellular matrix protein cytotactin. *J. Cell Biol.*, **119**, 663–678.
169. Prieto, A.L., Edelman, G.M. and Crossin, K.L. (1993) Multiple integrins mediate cell attachment to cytotactin/tenascin. *Proc. Natl. Acad. Sci. USA*, **90**, 10154–10158.
170. Probstmeier, R., Kühn, K. and Schachner, M. (1990) Fibroblast-derived J1 adhesion glycoproteins show binding properties to extracellular matrix constituents different from those of central nervous system origin. *J. Neurochem.*, **54**, 1016–1019.
171. Rakic, P. (1988) Specification of cerebral cortical areas. *Science*, **241**, 170–177.
172. Ranscht, B. and Dours, M.T. (1988) Sequence of contactin, a 130-kD glycoprotein concentrated in areas of interneuronal contact, defines a new member of the immunoglobulin supergene family in the nervous system. *J. Cell Biol.*, **107**, 1561–1573.
173. Rathjen, F.G. and Jessel, T.M. (1991) Glycoproteins that regulate the growth and guidance of vertebrate axons: domains and dynamics of the immunoglobulin/fibronectin type III subfamily. *Semin. Neurosci.*, **3**, 297–307.
174. Rathjen, F.G., Wolff, J.M. and Chiquet-Ehrismann, R. (1991) Restrictin: a chick neural extracellular matrix protein involved in cell attachment co-purifies with the cell recognition molecule F11. *Development*, **113**, 151–164.
175. Reichardt, L.F. and Tomaselli, K.J. (1991) Extracellular matrix molecules and their receptors: functions in neural development. *Annu. Rev. Neurosci.*, **14**, 531–570.
176. Rettig, W.J., Erickson, H.P., Albino, A.P. and Garinchesa, P. (1994) Induction of human tenascin (neuronectin) by growth factors and cytokines - cell type-specific signals and signalling pathways. *J. Cell Science*, **107**, 487–497.
177. Riou, J.-F., Shi, D.-L., Chiquet, M. and Boucaut, J.-C. (1990) Exogenous tenascin inhibits mesodermal cell migration during amphibian gastrulation. *Dev. Biol.*, **137**, 305–317.
178. Ruoslahti, E. (1988a) Structure and biology of proteoglycans. *Annu. Rev. Cell Biol.*, **4**, 229–255.
179. Ruoslahti, E. (1988b) Fibronectin and its receptors. *Annu. Rev. Biochem.*, **57**, 375–413.
180. Saga, Y., Tsukamoto, T., Jing, N., Kusakabe, M. and Sakakura, T. (1991) Murine tenascin: cDNA cloning, structure and temporal expression of isoforms. *Gene*, **104**, 177–185.

181. Saga, Y., Yagi, T., Ikawa, Y., Sakakura, T. and Aizawa, S. (1992) Mice develop normally without tenascin. *Genes&Devel.*, **6**, 1821–1831.
182. Sage, E.H. and Bornstein, P. (1991) Extracellular proteins that modulate cell-matrix interactions. *J. Biol. Chem.*, **266**, 14831–14834.
183. Salmivirta, M., Elenius, K., Vainio, S., Hofer, U., Chiquet-Ehrismann, R., Thesleff, I. and Jalkanen, M. (1991) Syndecan from embryonic tooth mesenchyme binds tenascin. *J. Biol. Chem.*, **266**, 7733–7739.
184. Sanes, J.R. (1989) Extracellular matrix molecules that influence neural development. *Annu. Rev. Neurosci.*, **12**, 491–516.
185. Schachner, M. (1991) Neural recognition molecules and their influence on cellular functions. In: The Nerve Growth Cone, eds. P.C. Letourneau, S.B. Kater, E.R. Macagno, New York: Raven, pp. 237–254.
186. Schnell, L., and Schwab, M.E. (1990) Axonal regeneration in the rat spinal cord produced by an antibody against myelin-associated neurite growth inhibitors. *Nature*, **343**, 269–272.
187. Schnell, L., Schneider, R., Kolbeck, R., Barde, Y.-A. and Schwab, M.E. (1994) Neurotrophin-3 enhances sprouting of corticospinal tract during development and after adult spinal cord lesion. *Nature*, **367**, 170–173.
188. Schwab, M.E. and Caroni, P. (1988) Oligodendrocytes and CNS myelin are nonpermissive substrates for neurite growth and fibroblast spreading in vitro. *J. Neurosci.*, **8**, 2381–2393.
189. Schwab, M.E., Kapfhammer, J. and Bandtlow, C.E. (1993) Inhibitors of neurite growth. *Annu. Rev. Neurosci.*, **16**, 565–595.
190. Sharifi, B.G., Lafleur, D.W., Pirola, C.J., Forrester, J.S. and Fagin, J.A. (1992) Angiotensin II regulates tenascin gene expression in vascular smooth muscle cells. *J. Biol. Chem.*, **267**, 23910–23915.
191. Sheppard, A.M., Hamilton, S.K. and Pearlman, A.L. (1991) Changes in the distribution of extracellular matrix components accompany early morphogenetic events of mammalian cortical development. *J. Neurosci.*, **11**, 3928–3942.
192. Siri, A., Carnemolla, B., Saginati, M., Leprini, A., Casari, G., Baralle, F. and Zardi, L. (1991) Human tenascin: primary structure, pre-mRNA splicing patterns and localization of the epitopes recognized by two monoclonal antibodies. *Nucleic Acid Res.*, **19**, 525–531.
193. Small, R., Riddle, P. and Raff, M.C. (1987). Evidence for migration of oligodendrocyte-type-2 astrocyte progenitor cells into the developing rat optic nerve. *Nature*, **328**, 155–157.
194. Snow, D.M., Steindler, D.A. and Silver, J. (1990a) Molecular and cellular characterization of the glial roof plate of the spinal cord and optic tectum: a possible role for a proteoglycan in the development of an axon barrier. *Dev. Biol.*, **138**, 359–376.
195. Snow, D.M., Lemmon, V., Carrino, D.A., Caplan, A.I. and Silver, J. (1990b) Sulfated proteoglycans in astroglial barriers inhibit neurite outgrowth in vitro. *Exp. Neurol.*, **109**, 111–130.
196. Snow, D.M. and Letourneau, P.C. (1992) Neurite outgrowth on a step gradient of chondroitin sulfate proteoglycan (CS-PG). *J. Neurobiol.*, **23**, 322–336.
197. Spring, J., Beck, K. and Chiquet-Ehrismann, R. (1989) Two contrary functions of tenascin: dissection of the active sites by recombinant tenascin fragments. *Cell*, **59**, 325–334.

198. Sriramarao, P., Mendler, M. and Bourdon, M.A. (1993) Endothelial cell attachment and spreading on human tenascin is mediated by $\alpha 2\beta 1$ and $\alpha v\beta 3$ integrins. *J. Cell Sci.*, **105**, 1001–1012.
199. Stahl, B., Müller, B., von Boxberg, I., Cox, E.C. and Bonhoeffer, F. (1990) Biochemical characterization of a putative axonal guidance molecule of the chick visual system. *Neuron*, **5**, 735–743.
200. Steindler, D.A. and Cooper, N.G.F. (1987) Glial and glycoconjugate boundaries during postnatal development of the central nervous system. *Dev. Brain Res.*, **36**, 27–38 (1987).
201. Steindler, D.A., Cooper, N.G.F., Faissner, A. and Schachner, M. (1989a) Boundaries defined by adhesion molecules during development of the cerebral cortex: the J1/tenascin glycoprotein in the mouse somatosensory cortical barrel field. *Dev. Biol.*, **131**, 243–260.
202. Steindler, D.A., Faissner, A. and Schachner, M. (1989b) Brain "Cordones": glial and adhesion molecule-transient boundaries define developing functional units. *Comments Dev. Neurobiol.*, **1**, 29–60.
203. Steindler, D.A., O'Brien, T.F., Laywell, E., Harrington, K., Faissner, A. and Schachner, M. (1990) Boundaries during normal and abnormal brain development: *in vivo* and *in vitro* studies of glia and glycoconjugates. *Exp. Neurol.*, **109**, 35–56.
204. Steindler, D.A. (1993) Glial boundaries in the developing nervous system. *Annu. Rev. Neurosci.*, **16**, 445–470.
205. Steindler, D.A., Settles, D., Erickson, H.P., Laywell, E.D., Yoshiki, A., Faissner, A. and Kusakabe, M. (1995) Tenascin knockout mice: barrels, boundary molecules, and glial scars. *J. Neurosci.*, **15**, 1971–1983.
206. Stern, C.D., Norris, W.E., Bronner-Fraser, M., Fraser, S.E., Carlson, G.J., Faissner, A., Keynes, R.J. and Schachner, M. (1989) Complex expression of J1/tenascin-related molecules in the sclerotome of the chick embryo. *Development*, **107**, 309–319.
207. Takeichi, M. (1991) Cadherin cell adhesion receptors as a morphogenetic regulator. *Science*, **251**, 1451–1455.
208. Tan, S.-S., Crossin, K.L., Hoffman, S. and Edelman, G.M. (1987) Asymmetric expression in somites of cytotactin and its proteoglycan ligand is correlated with neural crest cell distribution. *Proc. Natl. Acad. Sci. USA*, **84**, 7977–7981.
209. Taylor, H.C., Lightner, V.A., Beyer Jr., W.F., McCaslin, D., Briscoe, G. and Erickson, H.P. (1989) Biochemical and structural studies of tenascin/hexabrachion proteins. *J. Cell. Biochem.*, **41**, 71–90.
210. Taylor, J., Pesheva, P., and Schachner, M. (1993) The influence of janusin and tenascin on growth cone behaviour *in vitro*. *J. Neurosci. Res.*, **35**, 347–362.
211. Tessier-Lavigne, M., and Placzek, M. (1991) Target attraction: are developing axons guided by chemotropism? *Trends Neurosci.*, **14**: 303–310.
212. Tolbert, L.P. and Oland, L.A. (1990) Glial cells form boundaries for developing insect olfactory glomeruli. *Exp. Neurol.*, **109**, 19–28.
213. Tucker, R.P. (1991) The distribution of tenascin and its transcript in the developing avian central nervous system. *J. Exp. Zool.*, **259**, 78–91.
214. Tucker, R.P., Hammarback, J.A., Jenrath, D.A., Mackie, E.J. and Xu, Y. (1993) Tenascin expression in the mouse: in situ localization and induction in vitro by

bFGF. *J. Cell Sci.*, **104**, 69–76.

215. Tucker, R.P. (1994) Guest Editor of the Special Issue *The function of tenascin: hypotheses and current knowledge. Persp. Dev. Neurobiol.*, **2** (1), 1–132.
216. Unsicker, K., Reichert-Preibsch, H., Schmidt, R., Pettmann, B., Labourdette, G. and Sensenbrenner, M. (1987) Astroglial and fibroblast growth factors have neurotrophic functions for cultured peripheral and central nervous system neurons. *Proc. Natl. Acad. Sci USA*, **83**, 5459–5463.
217. Varnum-Finney, B., Venstrom, K., Muller, U., Kypta, R., Backus, C., Chiquet, M. and Reichardt, L.F. (1995) The integrin receptor $\alpha 8\beta 1$ mediates interactions of embryonic chick motor and sensory neurons with tenascin-C. *Neuron* **14**, 1213–1222.
218. Vaughan, L., Huber, S., Chiquet, M. and Winterhalter, K.H. (1987) A major, six-armed glycoprotein from embryonic cartilage. *EMBO J.*, **6**, 349–353.
219. Vaughan, L., Weber, P., D´Alessandri, L., Zisch, A.H. and Winterhalter, K.H. (1994) Tenascin-contactin/F11 interactions: a clue for a developmental role? *Persp. Dev. Neurobiol.*, **2**, 43–52.
220. Walter, J., Kern-Veits, B., Huf, J., Stolze, B. and Bonhoeffer, F. (1987a) Recognition of position-specific properties of tectal cell membranes by retinal axons *in vitro. Development*, **101**, 685–696.
221. Walter, J., Henke-Fahle, S. and Bonhoeffer, F. (1987b) Avoidance of posterior tectal membranes by temporal retinal axons. *Development*, **101**, 909–913.
222. Weber, P., Zimmermann, D.R., Winterhalter, K.H. and Vaughan, L. (1995) Tenascin-C binds heparin by its fibronectin type III domain Five. *J. Biol. Chem.*, **270**, 4619–4623.
223. Wehrle, B. and Chiquet, M. (1990) Tenascin is accumulated along developing peripheral nerves and allows neurite outgrowth in vitro. *Development*, **110**, 401–415.
224. Wehrle-Haller, B., Koch, M., Baumgartner, S., Spring, J. and Chiquet, M. (1991) Nerve-dependent and -independent tenascin expression in the developing chick limb bud. *Development*, **112**, 627–637.
225. Wehrle-Haller, B. and Chiquet, M. (1993) Dual function of tenascin: simultaneous promotion of neurite growth and inhibition of glial migration. *J. Cell Sci.*, **106**, 597–610.
226. Weller, A., Beck, S. and Ekblom P. (1991) Amino acid sequence of mouse tenascin and differential expression of two tenascin isoforms during embryogenesis. *J. Cell Biol.*, **112**, 355–362.
227. Williams, E.J., Furness, J., Walsh, F.S. and Doherty, P. (1994a) Characterization of the second messenger pathway underlying neurite outgrowth stimulated by FGF. *Development*, **120**, 1685–1693.
228. Williams, E.J., Furness, J., Walsh, F.S. and Doherty, P. (1994b) Activation of the FGF receptor underlies neurite outgrowth stimulated by L1, N-CAM, and N-Cadherin. *Neuron*, **13**, 583–594.
229. Wintergerst, E.S., Fuss, B. and Bartsch, U. (1993) Localization of janusin mRNA in the central nervous system of the developing and adult mouse. *Eur. J. Neurosci.*, **5**, 299–310.
230. Yavin, E., Gabai, A. and Gil S. (1991) Nerve growth factor mediates monosialoganglioside-induced release of fibronectin and J1/tenascin from C6 glioma cells. *J. Neurochem.*, **56**, 105–112.
231. Yokosaki, Y., Palmer, E.L., Prieto, A.L., Crossin, K.L., Bourdon, M.A., Pytela, R.

and Sheppard, D. (1994) The integrin alpha 9 beta 1 mediates cell attachment to a non-RGD site in the third fibronectin type III repeat of tenascin. *J. Biol. Chem.*, **269**, 26691–26696.

232. Yoshida, J., Wakabayashi, T., Okamoto, S., Kimura, S., Washizu, K., Kiyosawa, K. and Mokuno, K. (1994) Tenascin in cerebrospinal fluid is a useful biomarker for the diagnosis of brain tumour. *J. Neurol. Neursurg. Psychiatr.*, **57**, 1212–1215.
233. Zagzag, D., Friedlander, D.R., Miller, D.C., Dosik, J., Cangiarella, J., Kostianovsky, M., Cohen, H., Grumet, M. and Greco, M.A. (1995) Tenascin expression in astrocytomas correlates with angiogenesis. *Cancer Res.*, **55**, 907–914.
234. Zisch, A.H., D'Alessandri, L., Ranscht, B., Falchetto, R., Winterhalter, K.H. and Vaughan, L. (1992) Neuronal cell adhesion molecule contactin/F11 binds to tenascin via its immunoglobulin-like domains. *J. Cell Biol.*, **119**, 203–213.

4. TENASCIN IN DISEASE

BARBARA CARNEMOLLA, ANNALISA SIRI, LAURA BORSI
and LUCIANO ZARDI

*Cell Biology Laboratory Istituto Nazionale per la
Ricerca sul Cancro Genoa, Italy*

As discussed in the previous three chapters, tenascin (TN) is a high molecular weight glycoprotein of the extracellular matrix that is composed of several protein domains homologous to epidermal growth factor, fibronectin (FN) type III repeats and fibrinogen. The TN protein and mRNA have a restricted distribution in normal adult tissues. Most important for the present chapter is that the expression of TN in many tissues is upregulated in a variety of disease states, during wound healing and during regeneration. In particular, the large TN isoform is expressed in many neoplastic tissues at much greater levels than in their normal tissue counterparts. This suggests that alternative splicing and its regulation may allow for particular biological functions that contribute to the disease phenotype. Some of these functions have been discussed in Chapter 2 and are described in greater detail here with regard to the regulation of alternative splicing of TN and the biologically significant functions resulting from the inclusion of particular fibronectin type III domains.

REGULATION OF THE ALTERNATIVE SPLICING OF THE TENASCIN PRIMARY TRANSCRIPT

As discussed in Chapter 1, the different human TN isoforms, in which a group of eight FN-like repeats is included or omitted, are generated by alternative splicing of the pre-mRNA. It has been shown that as many as eight different mRNA forms, containing different numbers of FN-like repeat sequences, are generated by various splicing patterns of the human TN pre-mRNA (Gulcher *et al.*, 1989; Siri *et al.*, 1991; Sriramarao and Bourdon, 1993) (Figure 4.1).

This observation raises two main questions: (1) what are the functional differences of the various TN isoforms? and (2) what are the factors involved in the control of TN alternative splicing?

Recently, a number of reports have documented functional differences among the various TN isoforms. Murphy-Ullrich *et al.* (1991) showed that focal adhesion integrity is down regulated by the alternative spliced domain of human TN, while Chiquet-Ehrismann *et al.* (1991) demonstrated that, in chicken, the TN isoform lacking the alternatively spliced region binds more strongly to FN than the isoform in which this region is included. Furthermore,

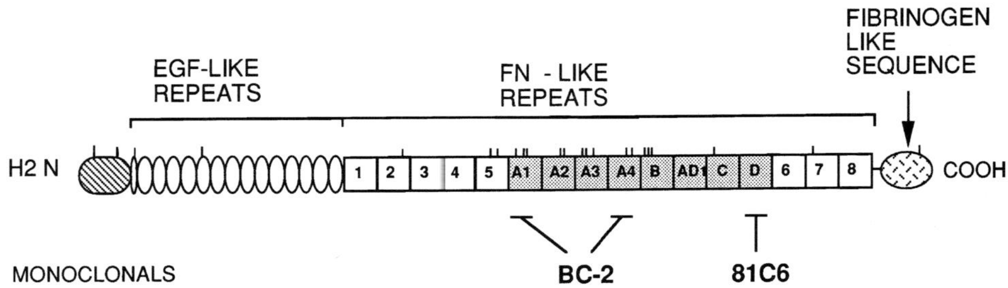

Figure 4.1 Model of the domain structure of a TN-C subunit. The FNIII repeats undergoing alternative splicing are indicated in grey. The regions recognized by the Mabs BC-2 and 81C6 are also indicated.

Kaplony *et al.* (1991) reported that the expression of the high molecular mass TN isoform is associated with corneal cell migration. Carnemolla *et al.* (1992) reported that malignantly transformed cultured human cells produce only the high molecular mass TN isoform, while their normal counterpart produces both isoforms of roughly 280 and 190 kDa. It is noteworthy that transformed cultured cells are unable to retain TN in the extracellular matrix, while normal human fibroblasts retain 60/90% of the produced TN. Moreover, in cultured normal fibroblasts, only the homohexameric TN molecules made up of the high molecular mass subunits are released into the culture medium, while the low molecular mass TN subunits are completely retained in the extracellular matrix (Carnemolla *et al.*, 1992). Recent reconstitution experiments using the two different TN isoforms have shown that while the small TN variant is able to form fibrils in the extracellular matrix of cultured cells, the large TN variant is not. Furthermore, it has recently been demonstrated that the TN molecules of the extracellular matrix have a heterohexameric structure (Siri *et al.*, submitted). This finding suggests that the assembly of TN in the extracellular matrix requires one or more small subunits in the TN molecule. It has also been reported that the two main TN isoforms bind to the cell surface through different receptors (Zisch *et al.*, 1992; Chung and Erickson, 1994; Vaughan *et al.*, 1994).

All available data on TN isoform distribution in different normal and pathological tissues (summarized in the following pages) imply that the large TN isoform is expressed at the onset of important cellular processes that entail active cell migration, proliferation or tissue remodelling, such as occur in wound healing, neoplasia and embryonic development. However, further studies are needed to clarify the possible biological functions of the different TN isoforms, also in light of the fact that nearly all studies to date have investigated only the two major isoforms: the small one, without the type III repeats from A to D (Fig. 4.1) and the large one, which includes the A–D repeats. In addition to these two major variants, others containing different numbers of FN III repeats have been described (Siri *et al.*, 1991).

It is worth noting that an altered alternative splicing in malignantly transformed cells is not limited to the TN pre-mRNA, the first observations in this regard being revealed through studies on the FN primary transcript (Castellani *et al.*, 1986; Borsi *et al.*, 1987, Zardi *et al.*, 1987; Carnemolla *et al.*, 1989). Furthermore, altered alternative splicing has also been observed in a number of primary transcripts including the pre-mRNA of protein kinase C in human colorectal cancer (Kuranami *et al.*, 1995) and that of the CD44 in a number of tumors (Tanabe and Saya, 1994). Thus, given that altered alternative splicing in tumors occurs frequently, it may constitute an important biological phenomenon and the products generated thereby may represent important targets for new therapeutic approaches.

In light of these preliminary studies on the functions of the different isoforms, clarifying the mechanisms involved in the regulation of TN alternative splicing becomes both intriguing and compulsory. Alternative pre-mRNA splicing is an important reversible mechanism of gene regulation which enables a single gene to encode multiple functionally different proteins. Although it is generally held that the pattern of alternative RNA splicing is only cell-type specific or developmentally regulated, its reversibility implies that it might be tuned by extracellular cues which enable cells to generate proteins with activities that allow adequate response to altered environmental conditions. For example, modifications of the TN isoform ratio induced by cytokines have been documented. Schwögler *et al.* (1992) reported modifications in the steady-state level of the different TN mRNAs induced by TGF-b in cultured rat smooth muscle cells. Tucker *et al.* (1993) reported that in 3T3 cells, b-FGF preferentially induced the accumulation of the large TN isoform. However, the TN primary transcript is not the only one showing a growth factor inducible alternative splicing pattern: Balza *et al.* (1988) and Borsi *et al.* (1990), studying the modulation of the alternative splicing of the fibronectin transcript induced by TGF-b, first revealed this behavior. Modulation of the alternative splicing of nontransmembrane phosphotyrosine phosphatase 1B has also been demonstrated to be cytokine inducible (Shifrin and Neel, 1993).

It has been observed that in breast lesions of varying nature, the high molecular mass TN isoform is a marker of stromal element proliferation (Borsi *et al.*, 1992). Furthermore, in experiments using cultured cells, it was demonstrated that the alternative splicing of the TN transcript is cell-cycle dependent and that its pattern of modification is a primary response to mitogenic stimulation (Borsi *et al.*, 1994). In fact, mitogenic stimulation of cultured fibroblasts using serum or cytokines leads to an early and striking modification in the steady-state levels of the two major TN mRNAs which have been shown to have equal instability. *De novo* protein synthesis is not necessary for this splicing modification. Similarly, changes both in synthesis and accumulation of the different TN isoforms are induced by mitogenic stimulation (Borsi *et al.*, 1994). In addition, in cultured normal human fibroblasts, small pH variations of the culture medium (from 7.2 to 6.9) strikingly modify the alternative splicing pattern of the TN primary transcript. Since such extracellular pH variations occur in many normal and pathological conditions, microenvironmental pH may be an important element for the regulation of RNA alternative splicing *in vivo* (Borsi *et al.*, 1995). Given that the modification of microenvironmental pH induces parallel modification of the steady-state intracellular pH in normal cells, such intracellular pH variations could affect the activity of a number of enzymes, and likewise the affinity of regulatory elements for RNA (Le Boeuf *et al.*, 1992; Mellergård *et al.*, 1994; Ou-Yang *et al.*, 1994; Katafuchi *et al.*, 1993).

The more acidic microenvironment of neoplastic with respect to normal tissues and the preferential accumulation of the large TN isoform in various types of cancer contrast with the above observation. This, and the fact that malignantly transformed or tumor-derived cultured cells produce only the large TN isoform (Borsi *et al.*, 1994), suggest that transformed cells do not respond to microenvironmental pHs as normal cells do. In fact, the SV-40 transformed fibroblast cell line WI-38-VA13 accumulate only the large TN mRNA, regardless of the pH of the media. Identical results were obtained with the melanoma cell line SK-MEL-28 (Borsi *et al.*, submitted). The lack of response to acidic environmental pHs of malignantly transformed cells could be explained by their greater ability, with respect to normal cells, to maintain a basic intracellular pH in an acidic environment. In fact, although neoplastic cells, at least in solid tumors, are surrounded by a more acidic extracellular fluid with respect to normal tissues, they often have a more basic intracellular pH than normal counterparts (Vaupel *et al.*, 1989). Furthermore, it has been reported that malignantly transformed cultured cells, as well as cells transformed by transfection with the oncogenes vErbA or c-H-ras, tolerate wider ranging culture medium pH variations than do non-transformed cells (Eagle, 1974; Fuerstenberg *et al.*, 1992; Kaplan and Boron, 1994). Very likely, environmental pH in normal cells controls TN alternative splicing, controlling in turn the cells' ability to respond to cytokines. In fact, at a pH below 7.0, normal cells respond neither to TGF-b nor to PDGF or EGF. As a consequence, extracellular pH may control the transition from the cell-cycle phase G0 to G1 (Borsi *et al.*, submitted). It is intriguing that, in turn, TN substrate controls the intracellular pH (Crossin, 1991; Krushel *et al.*, 1994) and, in this way, TN controls many cellular functions including TN expression, TN alternative splicing and cellular proliferation.

TN IN REACTIVE, HYPERPLASTIC AND NEOPLASTIC TISSUES

The breast

TN is found only in very small amounts in normal breast tissue, mainly localized in vessels and in a thin continuous rim surrounding ducts and acini (Howeedy *et al.*, 1990; Nicolò *et al.*, 1990). First data on increased TN expression in mammary tumors were documented by Chiquet-Ehrismann *et al.* (1986) and

Mackie *et al.* (1987) following studies on carcinogen-treated rats, virus-induced mammary tumors from mice and on a variety of human mammary gland lesions. According to their observations, TN is detectable in the stroma of malignant but not of benign tumors from all species, thus prompting consideration of TN as a possible stromal marker for epithelial malignancy in the mammary gland. However, further evidence has shown that the appearance of TN is not necessarily related to malignancy, but is rather a marker of stromal element proliferation (Borsi *et al.*, 1992). Increased TN expression is observed and clearly enhanced around hyperplastic ducts in all variants of mammary fibrocystic diseases. Fibroadenomas, benign Phylloid tumors, as well as *in situ* carcinomas, show an uneven TN periductal reaction. On the contrary, in all invasive carcinomas (ductal and lobular) the tumor stroma shows an extensive and strong reaction that is most intense at the tumor invasive edge and in septa around and within cell nests (Sakakura, 1983; Ferguson *et al.*, 1990; Gould *et al.*, 1990; Nicolò *et al.*, 1990; Koukoulis *et al.*, 1991; Natali *et al.*, 1991; Koukoulis *et al.*, 1993; Moch *et al.*, 1993).

Patterns of alternative splicing of the TN primary transcript in normal, hyperplastic and neoplastic breast tissues have also been examined. In all invasive breast carcinomas analyzed, the relative amount of TN mRNA in which the alternatively spliced region was included was about 10 times higher than in normal breast tissues. Similar findings were observed in Phylloides tumors and in those fibroadenomas which showed very high stromal cellularity. Western-blot analysis using different monoclonal antibodies confirmed the expression pattern detected by Northern blotting (Borsi *et al.*, 1992).

The Central Nervous System (CNS)

Different areas of normal adult cerebrum, cerebellum and spinal cord show little or no TN immunoreactivity in the white matter and meninges, while convincing, albeit focal and weak staining, is noted in the cerebral cortical matrix. Strong perineurial staining is observed in the large nerves (Koukoulis *et al.*, 1991). In intracranial abscesses and tuberculoma, strong TN staining is detected in the matrix and in the perivascular connective tissue of proliferating vessels. The positive reaction is enhanced in areas undergoing fibrous organization (Castellani *et al.*, in press). Bourdon *et al.* (1983) described TN expression and accumulation in glioblastoma, and several reports on human gliomas have confirmed this initial finding. Heterogeneous TN expression is observed in glioblastomas whose most constant feature is the marked staining of the extracellular stroma and around the walls of hyperplastic blood vessels. Consistent reaction is found to occur usually in the bundles of spindle-shaped cells, as well as in the cytoplasm of the cells arranged in palisades. Low-grade astrocytomas exhibit widespread staining both in the fibrillary background and in the connective stroma. Variability in both intensity and distribution of staining is more prominent in anaplastic astrocytomas. An enhanced positive reaction in the stroma of malignant astrocytomas is often observed (Koukoulis *et al.*, 1991; Higuchi *et al.*, 1993; Zagzag *et al.*, 1995; Castellani *et al.*, in press). In other brain tumors of different histotypes (neurinoma, ependymoma and meningioma), a positive reaction is always revealed in perivascular stroma. The strong association of TN and vascular hyperplasia, regardless of tumor grade, suggests that TN may play a crucial role in angiogenesis (Zagzag *et al.*, 1995). In intracranial carcinoma metastasis, the stroma reacts strongly with a diffuse positivity in the areas where epithelial cell clusters are intermingled within a compact connective stroma.

The liver

In the normal liver, reactivity along the sinusoids is reported. In hepatocellular carcinomas, increased stromal expression is observed with prominent localization in the interface between tumor and invaded normal tissues and within cancerous sinusoids. Furthermore, an inverse relationship between tumor differentiation and TN immunoreactivity is observed. Several hepatocellular carcinomas show a sharp contrast between the delicate staining of their own matrix and the marked reactivity in adjacent cirrhotic nodules (Koukoulis *et al.*, 1991; Natali *et al.*, 1991; Yamada *et al.*, 1992; Jaskiewicz *et al.*, 1993). In micronodular liver cirrhosis, an increase of extent and intensity of the perisinusoidal staining is noted. Variable extensive staining is detected in chronic active hepatitis (Koukoulis *et al.*, 1991).

The lung

In normal lung, strong staining is present in the smooth muscle of bronchi and bronchioles, while their basement membranes as well as the connective tissue of the alveolar septa show only a weak reaction. A stronger reaction in the subpleural connective tissue is reported. Variable strong staining for TN is reported in interstitial pneumonitis (Koukoulis *et al.*, 1991). An increased TN immunoreactivity is seen in the stroma of all carcinomas of various histologic types. Staining intensity is stronger in the stroma of squamous cell carcinomas than in the stroma of the other types of lung carcinomas. A portion of lung carcinomas also express intracytoplasmic TN immunoreactivity, suggesting that tumor cells may be able to synthesize TN. TN expression is not restricted to malignant neoplasms, but the presence of TN in non-neoplastic lung parenchyma and in some benign lung tumors has been reported (Koukoulis *et al.*, 1991; Natali *et al.*, 1991; Soini *et al.*, 1993). Moreover, lung carcinoma tissues show an increased expression of the large TN isoform compared to normal lung tissue (Oyama *et al.*, 1991). In interstitial pulmonary fibrosis, alveolar septal thickening and collapse is associated with enhancement of TN expression parallel to the capillary network (Koukoulis *et al.*, 1991).

Oral mucosa, salivary glands, stomach and colon

In normal buccal and palatal mucosa, in ventral tongue, in the floor of the mouth and in gingiva, immunoreaction for TN is described as a continuos thin, delicate line merely in the basement membrane region. Hyperkeratosis without dysplasia shows a distinct zone of enhanced TN immunoreactivity immediately below the epithelium. In dysplasias of various degree, enhancement of the stromal TN content may be observed, being most conspicuous in *in situ* carcinoma lesions. In most invasive carcinoma TN immunoreactivity is more intense, extending deeply into the underlying stroma with increased expression at the advancing edges of the tumor (Harada *et al.*, 1994; Tiitta *et al.*, 1994a).

In normal salivary glands, TN is merely found in periductal tissue and in blood vessels. In non-neoplastic tissue, TN is seen in the areas of basement membranes of the ductal epithelium. In pleomorphic adenomas, TN is seen in the stromal compartment and is more pronounced in the dense stromal areas and chondroid elements than in the myxoid areas. In Warthin's tumors, strong TN immunoreactivity may be observed in the basement membrane zone of the epithelial component. In adenoid cystic carcinomas, acinar cell tumors and mucoepidermoid carcinomas, TN shows stromal accumulation around the neoplastic lesion. No intracytoplasmic immunoreactivity is detectable in any of these cases (Soini *et al.*, 1992; Sunardhiwidyaputra and Vandamme 1993).

TN expression has been reported in the antrum and the body of the normal human stomach, immunoreactivity being detected only in the muscularis mucosae, while in the duodenum, a TN immunoreactive rim is seen underlying the surface of epithelial cells. Prominent TN immunoreaction is seen at the base of both benign and malignant ulcers, extending deep into underlying muscularis. Only severely dysplastic lesions display TN in the lamina propria in close association with capillaries. In early forms of diffuse gastric cancer, increased TN staining is seen in the lamina propria. In advanced forms, consistent TN expression is reported in the tumor stroma. A distinct TN reaction is seen surrounding invasive tumor cell nests of intestinal type gastric cancer. Thus, the fact that TN enhancement in the stomach is most conspicuous in invasive tumors and in ulcers suggests that TN is an important stromal component in malignant growth and in lesions undergoing active repair and remodelling. In some examples of moderate and severe chronic gastritis, a delicate TN-reactive line is seen to underscore the basement membrane of the epithelium (Tiitta *et al.*, 1994b).

Conflicting data have been reported on TN expression in normal adult colon: no immunoreactivity is described by Sakai *et al.* (1993) while other reports have described TN expression in the basement membrane of the superficial epithelium, muscularis mucosa and muscle layer (Oike *et al.*, 1990; Natali *et al.*, 1991; Hauptmann *et al.*, 1995). In neoplastic lesions, TN is described in the fibrous stroma surrounding the glandular epithelia. The neoplastic cells themselves are totally negative for TN expression. Both the tubular adenoma tissues and the superficial layer of well-differentiated adenocarcinomas are intensely reactive to TN antibodies, and the staining intensity increases as the adenomas became more atypical. The staining is most intense adjacent to the neoplastic epithelium and shows an uneven distribution in the stroma of the central part of the tumors (Hirota *et al.*, 1990; Sugawara *et al.*, 1991). Two major isoforms of TN with relative molecular masses of about 300 and 200 kDa are detected from colon cancer tissue (Sakai *et al.*, 1993; Hauptmann *et al.*, 1995).

The kidney

Immunohistochemical studies have shown a TN expression in normal adult kidney limited to the mesangium, focally in Bowman's capsule, the peritubular interstitium, in the medulla and in the pelvis (Natali *et al.*, 1991; Truong *et al.*, 1994). Weak and delicate staining is detected in oncocytomas and variably increased staining in carcinomas (Koukoulis *et al.*, 1991; Natali *et al.*, 1991). In glomerulopathies, regardless of the etiologies and morphologic types of glomerular disease (including lupus nephritis, post infectious glomerulonephritis, diabetic nephropathies, ischemia), whenever there is expansion of the extracellular matrix, be it in the mesangial, endocapillary, or extracapillary space, there is a concomitant and proportional increase in TN staining (Natali *et al.*, 1991; Assad *et al.*, 1993; Truong *et al.*, 1994). Furthermore, high levels of TN are seen in atherosclerotic plaques, particularly in the intima and subintima (Koukoulis *et al.*, 1991) and the TN isoform expressed to the greatest extent is the one containing all the type III repeats undergoing alternative splicing (Sane *et al.*, 1995). In autosomal dominant polycystic kidney disease, the strong increase of fibrogenesis is correlated with a strong expression of TN, with marked intensity adjacent to cyst-lining epithelium (Klingel *et al.*, 1993). TN is also strongly expressed in the interstitial and vascular compartment of acute and chronic renal rejection injury, suggesting that active repair and tissue remodelling occur in these sites (Gould *et al.*, 1992).

Prostate and urinary bladder

Normal adult prostate tissue reveals only focal, scant periglandular and stromal immunoreactivity around acini and ducts. The majority of cases of hyperplasia and intraepithelial neoplasia shows variable periglandular immunostaining. TN periglandular staining with diffuse stromal extension is noted with all grades of adenocarcinoma; however, the intensity is variable and appears to be unrelated to the histologic grade. Western blot analysis of prostate hyperplasia and carcinoma demonstrates the presence of both the small and large isoforms of TN (Natali *et al.*, 1991; Ibrahim *et al.*, 1993).

In normal adult urinary bladder, TN reaction is delicate and not extensive, while in flat *in situ* carcinomas, TN staining is stronger. In chronic cystitis, TN reactivity is enhanced particularly around capillary blood vessels, and in some cases, stronger than in flat *in situ* carcinomas (Tiitta *et al.*, 1993). In malignant neoplasias of urinary bladder, TN is found in the sub-urothelial lamina propria and in some urothelial basement membranes. The stroma of invasive tumors reacts strongly, the reaction being correlated with infiltrating grade (Tiitta *et al.*, 1993; Deen and Ball, 1994)

Endometrium and ovary

A very limited TN expression is described in normal adult endometrium, while uterine cervix, fallopian tubes and ovary are completely negative (Natali *et al.*, 1991). A weak expression of TN is reported in endometrium during the proliferative phase of the menstrual cycle and in hyperplasia. On the contrary, the secretory endometrium is negative. Invasive endometrial carcinomas display intense and diffuse staining around the carcinoma cells. The intensity of TN staining in endometrial carcinomas is not related to the degree of tumor differentiation. A similar expression is reported in uterine sarcomas, with TN distribution exclusively within the extracellular compartment (Vollmer *et al.*, 1990; Sasano *et al.*, 1993; Vollmer *et al.*, 1993). Finally, TN expression is detected in major septa and vessels of ovarian carcinomas (Natali *et al.*, 1991).

Lymphatic tissue

TN expression has been demonstrated in all peripheral lymphoid tissues, whereas it is barely detectable in the thymus and in the bone marrow. In reactive lymph nodes, TN is mainly found in T-dependent zones forming a variable reticula network corresponding to fibroblastic reticulum cells and blood vessel basal laminae; in B-dependent zones, TN is restricted to blood vessels. The TN network appears much thicker in chronically stimulated tissues and at sites of ongoing inflammation. In particular, a striking increase of TN is detected in sarcoid lymph nodes, as well as in myasthenic hyperplastic thymuses. In addition, TN can be accumulated in tissues involved in various types of lymphomas, including Hodgkin's disease and hairy cell leukemia (Liakka and Autio-Harmainen, 1992; Soini *et al.*, 1992; Chilosi *et al.*, 1993).

Soft tissues

Delicate and diffuse TN staining is observed in outer perichondrium, tendons, smooth muscles, perineurium and Schwann cells (Koukoulis *et al.*, 1991). TN expression is associated with Dupuytren's disease (Berndt *et al.*, 1994). An increased expression has been detected in soft tissue tumors (leiomyomas, leiomyomasarcomas, fibrosarcomas, malignant fibrous histiocytomas, malignant schwannomas, chondrosarcomas and osteogenic sarcomas). TN stromal staining is localized in small peri-intralesional septa, major septa and in blood vessels (Koukoulis *et al.*, 1991; Natali *et al.*, 1991)

The skin

In normal human skin sections, an intensive staining in the papillary derma and in the vessels has been reported (Nicolò *et al.,* 1990; Natali *et al.*, 1991; Shikata *et al.*, 1994), as well as in the ductal portion of the sweat glands, around the lower part of the hair follicle and in hair bulbs. Increased TN expression is documented in acne keloids and acne vulgaris and in hyperproliferative skin disease (Schalkwijk *et al.*, 1990; Schalkwijk *et al.*, 1991; Wood *et al.*, 1991; Knaggs *et al.*, 1993; Knaggs *et al.*, 1994). Increased reaction is described in stroma of basal, spinocellular, and squamous carcinomas (Stamp, 1989; Koukoulis *et al.*, 1991; Natali *et al.*, 1991). Tumors originating from sweat glands and hair follicles express TN around the tumor cell nests, while sebaceous gland tumors are negative for TN. Thus, TN expression in skin appendage tumors generally resembles that observed in corresponding normal tissue (Shikata *et al.*, 1994). In benign and malignant lesions of melanocyte lineage, TN displays a heterogeneous distribution, and a higher degree of expression is reported in lesions of greater dermal invasiveness. TN expression is described also in bland naevi and data reveal an increase of TN content from bland naevi through lesions with progressive atypia to invasive melanoma (Natali *et al.*, 1990; Koukoulis *et al.*, 1991; Tuominen and Kallioinen, 1994; Vanduinen *et al.*, 1994).

Articular cartilage

TN is weakly expressed in normal adult cartilage, while in samples from osteoarthritis (OA) and rheumatoid arthritis (RA), it is much more expressed. In particular, whereas in OA cartilage TN accumulates in the superficial layers and only in advanced lesions is it present in deeper zones, in RA cartilage it accumulates diffusely in the middle and deep zones (Cutolo *et al.*, 1992; Chevalier*et al.,* 1994). McCachren and Lightner (1992) have demonstrated that both TN protein and mRNA are increased in inflammatory synovitis and that interleukin-1 (a major mediator of tissue injury in inflammatory synovitis) can alter the extracellular matrix composition and deposition in primary synovial fibroblast cultures.

CLINICAL USE OF MONOCLONAL ANTIBODIES TO THE LARGE TN ISOFORM

The observations described above provide materials and models for studies on relevant biological problems such as the mechanisms regulating the alternative splicing and differential functions of the various TN isoforms.

Regardless of the biological significance of these observations, the large TN isoform, which is expressed in neoplastic tissues to a much greater extent than in normal tissues, may represent an important target for different clinical applications. Gliomas, for instance, have already benefited from clinical approaches using TN mAbs, due to two principal reasons: one, TN was detected more than ten years ago as a major component of these neoplasias (Bourdon *et al.*, 1983), and two, gliomas are usually well compartimentalized, are highly invasive, but in general do not metastasize and thus are suitable for direct intralesional therapy (Riva *et al.*, 1992).

Two mAbs, both specific for the large TN isoform, have thus far found clinical application. The first is a mouse anti-TN mAb (81C6) with binding for TN localized in the FNIII domain D (see Figure. 4.1) (Bourdon *et al.*, 1983). This murine anti-TN mAb, following iodination, has been shown to be an efficient localizing and therapeutic agent in both subcutaneous and intracranial human glioma xenograft models in athymic mice and rats (Bourdon *et al.*, 1984; Bullard *et al.*, 1986; Lee *et al.*, 1988a, 1988b). It has also been used for selective tumor localization in glioma patients (Schold *et al.*, 1993) and Phase I trials are currently in progress (He *et al.*, 1994). To reduce the human anti-mouse response, He *et al.* (1994) produced a mouse/human chimeric antibody (ch81C6) in which the variable region of mouse antibody is joined to the constant regions of human immunoglobulins. In this way, the chimeric mAb retains the specificity and affinity of the mouse hybridoma from which the variable region genes were isolated.

Numerous studies have been carried out using the mAb BC-2 which recognizes two epitopes within TN molecules, one on the FNIII A1 and one on FNIII A4 (Figure ™4.1) (Balza *et al.*, 1993). This peculiarity doubles the number of mAb molecules that bind to a given TN molecule. The mAb BC-2 has been used in a 3-step approach (Paganelli *et al.*, 1991) for the imaging of gliomas in single photon emission tomography (SPET), in which the targeting of the tumor by the mAb is achieved before the actual introduction of radioactivity. Injection of biotinylated mAb (first step) is followed by avidin administration (second step) to precipitate circulating biotinylated mAb and, at the same time, to target the tumor cells in order to home in adequately the subsequently administered radiolabeled biotin (third step). This technique has led to an improved target/non target ratio. A recent study examined twenty patients with histologically documented cerebral lesions, 18 of which were gliomas (Paganelli *et al.*, 1994). In many lesions, immunohistochemical BC-2 staining of tissue samples obtained from open surgery or stereotactic biopsy revealed that TN was abundantly present in all glioblastomas and grade III astrocytomas. Using BC-2 in antibody-guided SPET, tumors were detected in 15 out of 18 glioma patients. Two of the non detected tumors were devoid of TN (oligodendrogliomas). It is noteworthy that in 3 out of the 15 positive patients, computerized tomography and/or magnetic resonance imaging scans had been inconclusive for the diagnosis of malignancy. No radioactivity uptake was recorded in the edema and scar tissue surrounding the tumor. The results suggest that the 3-step immuno scintigraphy using the mAb BC-2 is a safe practice since no toxicity nor alteration of renal and hepatic functions were observed. Moreover, in addition to being a useful reagent for the diagnosis of such lesions, BC-2 may be valuable in the follow-up of glioma patients in whom the presence of TN had been previously assessed by immunohistochemistry. Figure 4.2 shows a SPET imaging of a case of grade III astrocytoma using the mAb BC-2 in the three step method.

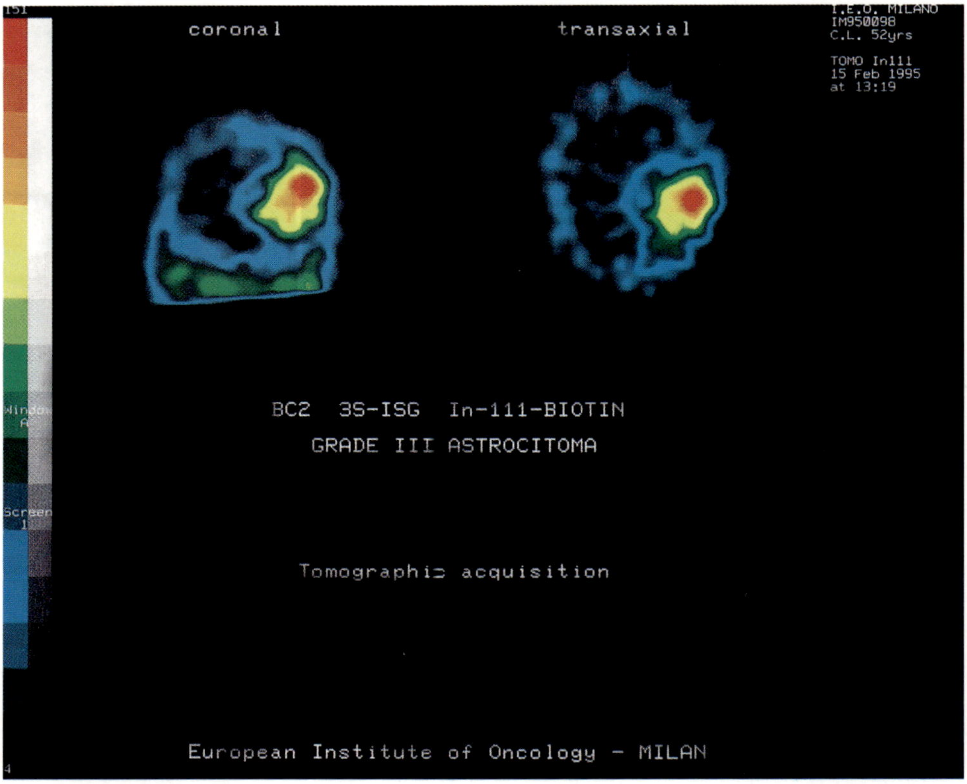

Figure 4.2 Tumor localization in a patient with grade III astrocytoma, using the BC-2-guided three-step SPET imaging, 1h after ^{111}In-biotin administration. Both images show an area of hyperaccumulation corresponding to the neoplastic lesion.
(The photo was kindly provided by Dr. G. Paganelli, European Institute of Oncology, Milan, Italy)

The mAb BC-2 has also been used in the radioimmunotherapy (RIT) of 48 patients with glioma, 28 of whom had recurrent tumors and 20 of whom were patients with newly diagnosed tumors. Radiolabeled BC-2 was given intralesionally to all patients, after previous assessment of TN expression in neoplastic tissue, at a volume which did not exceed 1.5 ml to avoid spikes in intracranial pressure. When possible, multiple cycles of intralesional treatment were given (two patients received 6 cycles) with an average mAb dose/cycle

Table 4.1 OBJECTIVE RESPONSE (WHO CRITERIA) TO RADIOIMMUNOTHERAPY OF 48 GLIOMA PATIENTS

response	*patients*	%	*time*
PD	19	39.5	6 1
SD	11	23	10
PR	6	12.5	10 4
CR	12	25	20 2

PD: Progressive disease; SD: stabilization of disease; PR: partial remission; CR: complete remission.
%: percentage of 48 patients experiencing the response. Time: median value of the response duration (in months).
Data kindly provided by Dr. P. Riva, in press on "Tumor Targeting"

of 3.4 mg, and a mean ^{131}I amount/cycle of MBq 1335.7 (Sturiale *et al.*, in press). Table 4.1 reports the objective response, established according to WHO criteria, to RIT and illustrates that a significant number of patients (12 out of 48) is still apparently free of disease after an average of more than 20 months from treatment. After four years, the median survival of the entire group (more than 50% of the patients had recurrent or bulky disease and very short life expectancy) is 18 months after RIT versus 12 months after conventional therapies. Even in those cases in which the treatment did not succeed in controlling the disease, the quality of life of the patients was improved (Sturiale *et al.*, in press). Thus, this therapeutic approach has proven capable of achieving, in some particular cases, the control of malignant glioma, which represents an otherwise intractable disease. At present, promising radioimmunotherapy trials are being carried out using the isotope ^{90}Y (Yttrium-90) to label biotin for the 3-step procedure (Paganelli, personal communication; Epenetos, personal communication).

Davico Bonino *et al.* (in press) have recently reported the production of a bispecific mAb combining the hybridoma BC-2 and the hybridoma CBT3G, to create a mAb specific for CD3. The resulting hybrid hybridomas were selected on the basis of the double specificity for CD3 and TN. This bi-mAb (CBT3G X BC-2) is able to simultaneously bind the two target antigens while maintaining the same reactivity as the parental BC-2 mAb in terms of distribution, homogeneity and intensity of staining in the tissues tested. The mitogenic activity of the CBT3G X BC-2 biMab on peripheral blood mononuclear cells (PBMC) is similar to that revealed by the parental anti-CD3 mAb. Furthermore, the hybrid molecule is also able to induce TNF-a gene expression in activated PBMC. Furthermore, the CBT3G X BC-2 is able to induce a significant cytotoxic activity against TN positive cultured tumor cells.

In conclusion, human-mouse chimeric anti-TN mAbs, bispecific mAbs and the use of more cytotoxic radioisotopes seem to be very promising clinical tools for glioblastoma that constitutes, to date, a cancer with an extremely severe prognosis.

REFERENCES

1. Assad, L., Schwartz, M.M., Virtanen, I. and Gould, V.E. (1993) Immunolocalization of tenascin and cellular fibronectins in diverse glomerulopathies. *Virchows Archiv.*, **63**, 307–316.
2. Berndt, A., Kosmehl, H., Katenkamp, D. and Tauchmann, V. (1994) Appearance of the myofibroblastic phenotype in Dupuytren's disease is associated with a fibronectin, laminin, collagen type IV and tenascin extracellular matrix. *Pathobiol.*, **62**, 55–58.
3. Balza, E., Borsi, L., Allemanni, G. and Zardi, L.(1988) Transforming growth factor-b regulates the levels of different fibronectin isoforms in normal human cultured fibroblasts. *FEBS Lett.*, **228**, 42–44.
4. Balza, E., Siri, A., Ponassi, M., Caocci, F., Linnala, A., Virtanen, I. and Zardi, L. (1993). Production and characterization of monoclonal antibodies specific for different epitopes of human tenascin. *FEBS Letters*, **332**, 39–43.
5. Borsi, L., Carnemolla, B., Castellani, P., Rosellini, C., Vecchio, D., Allemanni, G., Chang, S.E., Taylor-Papadimitriou, J., Pande, H. and Zardi, L. (1987) Monoclonal antibodies in the analysis of fibronectin isoforms generated by alternative splicing of mRNA precursors in normal and transformed human cells. *J. Cell Biol.*, **104**, 595–600.
6. Borsi, L., Castellani, P., Risso, A.M., Leprini, A. and Zardi, L. (1990) Transforming growth factor-b regulates the splicing pattern of fibronectin messenger RNA precursor. *FEBS L.*, **261**, 175–178.
7. Borsi, L., Carnemolla, B., Nicolò, G., Spina, B., Tanara, G., and Zardi, L. (1992) Expression of different tenascin isoforms in normal, hyperplastic and neoplastic human breast tissues. *Int. J. Cancer*, **52**, 688–692.
8. Borsi, L., Balza, E., Castellani, P., Carnemolla, B., Ponassi, M., Querzé, G., and Zardi L. (1994) Cell-cycle dependent alternative splicing of the tenascin primary transcript. *Cell Adhes. Commun.*, **1**, 307–317.
9. Borsi, L., Balza, E., Gaggero, B., Allemanni, G. and Zardi, L. (1995) The alternative splicing pattern of the tenascin-C pre-mRNA is controlled by the extracellar pH. *J. Biol. Chem.*, **11**, 6243–6245.
10. Bourdon, M.A., Wikstrand, C.J., Furthmayr, H., Matthews, T.J. and Bigner, D.D. (1983) Human glioma-mesenchymal extracellular matrix antigen defined by monoclonal antibody. *Cancer Res.*, **43**, 2796–2805.
11. Bourdon, M.A., Coleman, R.E., Blasberg, R.G., Groothius, D.R. and Bigner, D.D. (1984) Monoclonal antibody localization in subcutaneous and intracranial human glioma xenografts: paired-label and imaging analysis. *Anticancer Res.*, **4**, 133–140.
12. Bullard, D.E., Adams, C.J., Coleman, R.E., and Bigner, D.D. (1986) In vivo imaging of intracranial human glioma xenograftscomparing specific with nonspecific radiolabeled manoclonal antibodies. *J. Neurosurg.*, **64**, 257–262.
13. Carnemolla, B., Balza, E., Siri, A., Zardi, L., Nicotra, M.R., Bigotti, A. and Natali,. P.G. (1989) A tumor-associated fibronectin isoform generated by alternative splicing of messenger RNA precursors. *J. Cell Biol.*, **108**, 1139–1148.
14. Carnemolla, B., Borsi, L., Bannikov, G., Troyanovsky, S., and Zardi, L. (1992) Comparison of human tenascin expression in normal, SV-40 transformed and tumor derived cell lines. *Eur. J. Biochem.*, **205**, 561–567.

15. Castellani, P., Siri, A., Rosellini, C., Infusini, E., Borsi, L. and Zardi, L. (1986). Transformed human cells release different fibronectin variants than do normal cells. *J. Cell Biol.*, **103**, 1671–1677.
16. Castellani P., Dorcaratto A., Siri A., Zardi, L. and Viale G.L. (1995) Tenascin distribution in human brain tumors. *Acta Neuroch.*, in press.
17. Chevalier, X., Groult, N., Largetpiet, B., Zardi, L. and Hornebeck, W. (1994) Tenascin distribution in articular cartilage from normal subjects and from patients with osteoarthritis and rheumatoid arthritis. *Arthritis Rheum.*, **37**, 1013–1022
18. Chiquet-Ehrismann R., Mackie E.J., Pearson C.A. and Sakakura, T. (1986) Tenascin: an extracellular matrix protein involved in tissue interactions during fetal development and oncogenesis. *Cell*, **47**, 131–139.
19. Chiquet-Ehrismann, R., Matsuoka, Y., Hofer, U., Spring, J., Bernasconi, C. and Chiquet, M. (1991). Tenascin variants: differential binding to fibronectin and distinct distribution in cell cultures and tissues. *Cell Regul.*, **2,** 927–938.
20. Chilosi, M., Lestani, M., Benedetti, A., Montagna, L., Pedron, S., Scarpa A., Menestrina, F., Hirohashi, S., Pizzolo, G. and Semenzato, G. (1993) Constitutive expression of tenascin in T-dependent zones of human limphoid tissues. *Am. J. Pathol.*, **143**, 1348–1355.
21. Chung, C.Y. and Erickson, H.P. (1994) Cell surface annexin II is a high affinity receptor for the alternatively spliced segment of tenascin-C. *J. Cell Biol.*, **126**, 539–548.
22. Crossin, K.L. (1991) Cytotactin binding: inhibition of stimulated proliferation and intracellular alkalinization in fibroblasts. *Proc. Natl. Acad. Sci. USA*, **88**, 11403–11407.
23. Cutolo, M., Picasso, M., Ponassi, M., Sun, M.Z. and Balza, E. (1992) Tenascin and fibronectin distribution in human normal and pathological synovium. *J. Rheumatol.*, **19**, 1439–1447.
24. Davico Bonino, L., De Monte, L.B., Spagnoli, G.C., Vola, R., Mariani, M., Barone, D., Moro, A.M., Riva, P., Nicotra, M.R., Natali, P.G. and Malavasi, F. Bispecific monoclonal antibody anti-CD3 X anti-tenascin: an immunotherapeutic agent for human glioma. *Int. Jour. Cancer*, in press.
25. Deen, S. and Ball, R.Y. (1994) Basement membrane and extracellular interstitial matrix components in bladder neoplasia — Evidence of angiogenesis. *Histopathol.*, **25**, 475–481.
26. Eagle, H. (1974) Some effects of environmental pH on cellular metabolism and function. In: *Control of proliferation in animal cells* (Eds: B. Clarkson and R. Baserga) 1–11, Cold Spring Harbor Laboratories.
27. Ferguson, J.E., Schor, A.M., Howell, A. and Ferguson, M.W. (1990) Tenascin distribution in the normal human breast is altered during the menstrual cycle and in carcinoma. *Differentiation*, **42**, 199–207.
28. Fuerstenberg, S., Leitner, I., Schroeder, C., Schwarz, H., Vennstrom, B. and Beug, H. (1992) Transcriptional repression of band 3 and CAII in v-erbA transformed erythroblasts accounts for an important part of the leukaemic phenotype. *EMBO J.*, **11**, 3355–3365.
29. Gould, V.E., Koukoulis, G.K. and Virtanen, I. (1990) Extracellular matrix proteins and their receptors in the normal, hyperplastic and neoplastic breast. *Cell Diff. Dev.*, **32**, 409–416.

30. Gould, V.E., Martinez-Lacabe, V., Virtanen, I., Sahlin, K.M. and Schwartz, M.M. (1992) Differential distribution of tenascin and cellular fibronectins in acute and chronic renal allograft rejection. *Lab. Invest.*, **67**, 71–76.
31. Gulcher, J.R., Nies, D.E., Marton, L.S. and Stefanssonn, K. (1989). An alternatively spliced region of the human hexabrachion contains a repeat of potential N-glycosylation sites. *Proc. Natl. Acad. Sci. USA*, **86**, 1588–1592.
32. Harada, T., Shinohara, M., Nakamura, S. and Oka, M. (1994) An immunohistochemical study of the extracellular matrix in oral squamous cell carcinoma and its association with invasive and metastatic potential. *Virchows Archiv.*, **424**, 257–266.
33. Hauptmann, S., Zardi, L., Siri, A., Carnemolla, B., Borsi, L., Castelucci, M., Klosterhalfen, B., Hartung, P., Weis, J., Stöcker, G., Haubeck, H.D. and Kirkpatrick, C.J. Extracellular Matrix Proteins in Colorectal Carcinomas. Expression of Tenascin and Fibronectin Isoforms (1995) *Lab. Invest.*, **73**, 172–182.
34. He, X., Archer, G.E., Wikstrand, C.J., Morrison, S.L., Zalutsky, M.R., Bigner, D.D. and Batra, S.K. (1994) Generation and characterization of a mouse/human chimeric antibody directed against extracellular matrix protein tenascin. *J. Neuroimmunol.*, **52**, 127–137.
35. Higuchi, M., Ohnishi, T., Arita, N., Hiraga, S. and Hayakawa, T. (1993) Expression of tenascin in human gliomas — Its relation to histological malignancy, tumor dedifferentiation and angiogenesis. *Acta Neuropathol.*, **85**, 481–487.
36. Hirota, N., Sakai, T., Kanai, N., Yamada, S., Aonuma, T., Yokoyama, T. and Sakakura, T. (1990) Immunohistochemical demonstration of tenascin in colonic neoplasms of the humans and rats. *J. Cancer Res. Clin. Oncol.*, **116**, 269.
37. Howeedy, A.A., Virtanen, I., Laitinen, L., Gould, N.S., Koukoulis, G.K. and Gould, V.E. (1990) Differential distribution of tenascin in the normal, hyperplastic, and neoplastic breast. *Lab. Invest.*, **63**, 798–806.
38. Ibrahim, S.N., Lightner, V.A., Ventimiglia, J.B.,Ibrahim, G.K., Walther, P.J. Bigner, D.D. and Humphrey, P.A. (1993) Tenascin expression in prostatic hyperplasia, intraepithelial neoplasia, and carcinoma. *Human Path.*, **24**, 982–989.
39. Jaskiewicz, K., Chasen, M.R. and Robson, S.C. (1993) Differential expression of extracellular matrix proteins and integrins in hepatocellular carcinoma and chronic liver disease. *Anticancer Res.*, **13**, 2229–2238.
40. Kaplan, D.L. and Boron, W.F. (1994) Long-term expression of c-H-ras stimulates Na-H and Na+-dependent Cl-HCO_3 exchange in NIH-3T3 fibroblasts. *J. Biol. Chem.*, **269**, 4116–4124.
41. Kaplony, A., Zimmermann, D.R. Fisher, R.W., Imhof, B.A., Odermatt, B.F., Winterhalter, K.H. and Vaughan, L. (1991). Tenascin Mr, 220.000 isoform expression correlates with cornea cell migration. *Development*, **112**, 605–614.
42. Katafuchi, T., Hagiwara, H., Ito, T. and Hirose, S. (1993) A dramatic pH-dependent alteration in ANP receptor density: a note for using cultured cells. *Am. J. Physiol.*, **264**, C1345–C1349.
43. Klingel, R., Ramadori, G., Schuppan, D., Knittel, T., Zumbuschenfelde, K.H.M. and Kohler, H. (1993) Coexpression of extracellular glycoproteins undulin and tenascin in human autosomal dominant plycystic kidney disease. *Nephron.*, **65**, 111–118.
44. Knaggs, H.E., Layton, A.M., Jutley, J. and Cunliffe, W.J. (1993) Expression of tenascin in acne keloids. *J. Invest. Dermatol.*, **100**, 552.

45. Knaggs, H.E., Layton, A.M., Morris, C., Wood, E.J., Holland, D.B. and Cunliffe, W.J. (1994) Investigation of the expression of the extracellular matrix glycoproteins tenascin and fibronectin during acne vulgaris. *Brit. J. Dermatol.*, **130**, 576–582.
46. Koukoulis, G.K., Gould, V.E., Bhattacharyya, A., Gould, J.E., Howeedy, A.A. and Virtanen, I. (1991) Tenascin in normal, reactive, hyperplastic and neoplastic tissue biologic and pathologic implications. *Human Pathol.*, **22**, 636–643.
47. Koukoulis, G.K., Howeedy, A.A., Korhonen, M., Virtanen, I. and Gould, V.E. (1993) Distribution of tenascin, cellular fibronectins and integrins in the normal, hyperplastic and neoplastic breast. *J. Submicrosc. Cytol. Pathol.*, **25** 285–295.
48. Krushel, L.A., Prieto, A.L., Edelman, G.M. and Crossin, K.L. (1994) Differential effects of citotactin/tenascin fusion proteins on intracellular pH and cell morphology. *J. Cell. Physiol.*, **161**, 508–518.
49. Kuranami, M., Powell, C.T., Hug, H., Zeng, Z., Cohen, A.M. and Guillem, J.G. (1995) Differential expression of protein Kinase C isoforms in human colorectal cancers. *J. Surg. Res.*, **58**, 233–239.
50. LeBoeuf, R.A., Lin, P., Kerckaert, G. and Gruenstein, E. (1992) Intracellular acidification is associatyed with enhanced morphological transformation in Syrian hamster embryo cells. *Cancer Res.*, **52**, 144–148.
51. Lee, Y.-S., Bullard, D.E., Zalutsky, M.R., Coleman, R.E., Wikstrand, C.J., Friedman, H.S., Colapinto E.V. and Bigner, D.D. (1988a) Therapeutic efficacyof antiglioma mesenchymal extracellular matrix ^{131}I-radiolabelmed murine monoclonal antibody in a human glioma xenograft model. *Cancer Res.*, **48**, 559–566.
52. Lee, Y.-S., Bullard, D.E., Humphrey, P.A., Colapinto E.V., Friedman, H.S., Zalutsky, M.R., Coleman, R.E. and Bigner, D.D. (1988b) Treatment of intracranial human glioma xenografts with ^{131}I-labeled anti-tenascin monoclonal antibody 81C6. *Cancer Res.*, **48**, 2904–2910.
53. Liakka, K.A. and Autio-Harmainen, H.I. (1992) Distribution of the extracellular matrix proteins tenascin, fibronectin, and vitronectin in fetal, infant, and adult human spleens. *J. Histochem. Cytochem.*, **40**, 1203–1210.
54. Mackie, E.J., Chiquet-Ehrismann, R., Adams Pearson, C., Inaguma, Y., Taya, K., Kawarada, Y. and Sakakura, T. (1987) Tenascin is a stromal marker for epithelial malignancy in the mammary gland. *Proc. Natl. Acad. Sci. USA*, **84**, 4621–4625.
55. McCachren, S.S., and Lightner, V.A. (1992) Expression of human tenascin in synovitis and its regulation by interleukin-1. *Arthritis Rheum.*, **35**, 1185–1196.
56. Mellergård, P., Ou-Yang, Y. and Siesjö, B.K. (1994) Relationship between intra- and extracellular pH in primary cultures of rat astrocytes; *Am J. Physiol.*, **267**, C581–C589.
57. Moch, H., Torhorst, J., Durmuller, U., Feichter, G.E., Sauter, G. and Gudat, F. (1993) Comparative analysis of the expression of tenascin and established prognostic factors in human breast cancer. *Pathol. Res. Practice*, **189**, 510–514.
58. Murphy-Ullrich, J.E., Lightner, V.A., Aukhil, I., Yan, Y.Z., Erickson, H.P, and Hook, M. (1991) Focal adhesion integrity is downregulated by the alternatively spliced domain of human tenascin. *J. Cell Biol.*, **115**, 1127–1130.
59. Natali, P.G., Nicotra, M.R., Bartolazzi, A., Mottolese, M., Coscia, N., Bigotti, A. and Zardi, L. (1990) Expression and production of tenascin in benign and malignant lesions of melanocytes lineage. *Int. J. Cancer*, **46**, 586–590.

60. Natali, P.G., Nicotra, M.R., Bigotti, A., Botti, C., Castellani, P., Risso, A.M. and Zardi, L. (1991) Comparative analysis of the expression of the extracellular matrix protein tenascin in normal human fetal, adult and tumor tissues. *Int. J. Cancer*, **47**, 811–816.
61. Nicolò, G., Salvi, S., Oliveri, G. Borsi, L., Castellani, P. and Zardi, L. (1990) Expression of tenascin and of the ED-B containing oncofetal fibronectin isoform in human cancer. *Cell Diff. Dev.*, **32**, 401–408.
62. Oike, Y., Hiraiwa, H., Kawakatsu, H., Nishikai, M., Okinaka, T., Suzuki, T., Okada, A.,Yatani, R. and Sakakura,T. (1990) Isolation and characterization of human fibroblast tenascin. An extracellular matrix glycoprotein of interest for development studies. *Int. J. Dev. Biol.*, **34**, 309–317.
63. Ou-Yang, Y., Kristiàn, T., Mellergard, P. and Siesjö, B.K. (1994) The influence of pH on glutamate- and depolarization-induced increases of intracellular calcium concentration in cortical neurons in primary culture. *Brain Res.*, **646**, 65–72.
64. Oyama, F., Hirohashi, S., Shimosato, Y., Titani, K. and Sekiguchi, K. (1991) Qualitative and quantitative changes of human tenascin expression in transformed lung fibroblast and lung tumor tissues: comparison with fibronectin. *Cancer Res.*, **51**, 4876–4881.
65. Paganelli, G., Magnani, P., Zito, F. Villa, E., Stella, M., Lopalco, L., Siccardi, A.G. and Fazio, F. (1991) Three-step monoclonal antibody tumor targeting in carcinoembryonic antigen patients. *Cancer Res.*, **51**, 5960–5966.
66. Paganelli, G., Magnani, P., Zito, F., Lucignani, G., Sudati, F., Truci, G., Motti, E., Terreni, M., Pollo, B., Giovanelli, M., Canal, N., Scotti, G., Comi, G., Koch, P., Maecke, H.R. and Fazio, F. (1994) Pre-targeted immunodetection in glioma patients: tumour localization and single-photon emission tomography imaging of [^{99m}Tc]PnAO-biotin. *Eur. J. Nucl. Med.*, **21**, 314–321.
67. Riva, P., Arista, A., Sturiale, C., Moscatelli, G., Tison, V., Mariani, M., Seccamanni, E., Lazzari, S., Fagioli, L., Franceschi, G., Sarti, G., Riva, N., Natali, P.G., Zardi, L. and Scassellati, G.A. (1992) Treatment of intracranial human glioblastoma by direct intratumoral administration of 131I-labeled anti-tenascin monoclonal antibody BC-2. *Int. J. Cancer*, **51**, 7–13.
68. Sakai, T., Kawakatsu, H., Hirota, N., Yokoyama, T., Sakakura, T. and Saito, M. (1993) Specific expression of tenascin in human colonic neoplasms. *Brit. J. Cancer*, **67**, 1058–1064.
69. Sakakura T. (1983) Epithelial-mesenchymal interactions in mammary gland development and its perturbation in relation to tumorigenesis. In: *Understanding Breast Cancer. Clinical and Laboratory Concepts.* (eds. M.A. Rich, J.C. Hager, P. Furamanski) 261–284, New York, NY, Marcel Dekker.
70. Sane, D.C., Yang, T.F., Kutcher, M.A. and Braden, G.A. (1995) Expression of fibronectin (FN) and tenascin (TN) splice variants in human coronary-artery atherosclerosis (A) and restenosis (R). *Faseb J.*, **9**, 608.
71. Sasano, H., Nagura, H., Watanabe, K, Ito, K., Tsuiki, A., Sato, S., Yajima, A., Kusakabe, M. and Sakakura, T. (1993) Tenascin expression in normal and abnormal human endometrium. *Modern Pathol.*, **6**, 323–326.
72. Schalkwijk, J., Mackie, E., Oosterling, B., Perret, C., Koopman, R. and Happle, R. (1990) Tenascin expression in hyperproliferative skin diseases. *Arch. Dermatol. Res.*, **281**, 566.

73. Schalkwijk, J., Van Vlijmen, I., Oosterling, B., Perret C., Koopman, R., Van den Born, J. and Makie, E.J. (1991) Tenascin expression in hyperproliferative skin disease. *Brit. J. Dermatol.*, **124**, 13–20.
74. Schold, S.C. Jr., Zalutsky, M.R., Coleman, R.E., Glantz, M.J., Friedman, A.H., Jaszczak, R.J., Bigner, S.H. and Bigner, D.D. (1993) Distribution and dosimetry of I-123-labeled monoclonal antibody 81C6 in patients with anaplastic glioma. *Invest. Radiol.*, **28**, 488–496.
75. Schwögler, S., Odenthal, M., Meyer zum Büschenfelde, K.-H. and Ramadori, G. (1992). Alternative splicing products of the tenascin gene distinguish rat liver fat storing cells from arterial smooth muscle cells and skin fibroblasts. *Biochem. Biophys. Res. Commun.*, **185**, 768–775.
76. Shifrin, V.I. and Neel, B.G. (1993) Growth factor-inducible alternative splicing of non transmembrane phosphotyrosine phosphatase PTP-1B pre-mRNA. *J. Biol. Chem.*, **268**, 25376–25384.
77. Shikata, N., Oyaizu, T., Andachi, H. and Tsubura, A. (1994) Tenascin expression in normal human adult skin and skin appendege tumors. *Virchows Archiv.*, **424**, 511–516.
78. Siri, A., Carnemolla, B., Saginati, M., Leprini, A., Casari, G., Baralle, F. and Zardi L. (1991). Human tenascin: primary structure, pre-mRNA splicing patterns and localization of the epitopes recognized by two monoclonal antibodies. *Nucl. Acids Res.*, **19**, 525–531.
79. Soini, Y., Alavaikko, M., Lehto, V.P. and Virtanen, I. (1992a) Tenascin in reactive lymph nodes and in malignant lymphomas. *Pathol. Res. Practice*, **188**, 1078–1082.
80. Soini, Y., Virtanen, P.I. and Lehto, V.P. (1992b) Tenascin in salivary gland tumours. *Virchows Archiv.*, **421**, 217–222.
81. Soini, Y., Paakko, P., Nuorva, K., Kamel, D., Linnala, A., Virtanen, I. and Lehto, V.P. (1993) Tenascin Immunoreactivity in Lung Tumors. *Am. J. Clin. Pathol.*, **100**, 145–150.
82. Sriramarao, P. and Bourdon, M.A. (1993) A novel tenascin type III repeat is part of a complex of tenascin mRNA alternative splices. *Nucl. Acids Res.*, **21**, 163–168.
83. Stamp, G.W.H. (1989) Tenascin distribution in basal cell carcinomas. *J. Pathol.*, **159**, 225–229.
84. Sturiale, C., Arista, A., Lazzari, S., Frattarelli, M., Donati, R., Sarti, G., Campori, F., Riva, N., Casi, M., Gentile, R., Rossitti, R., and Riva, P. Intralesional radioimmunotherapy of malignant gliomas: clinical experiences in recurrent and primary tumours. *Tumor Targeting*, in press.
85. Sugawara, I., Hirakoshi, J., Masunaga, A., Itoyama, S. and Sakakura, T. (1991) Reduced tenascin expression in colonic carcinoma with linfogenous metastasis. *Inv. Metast.*, **11**, 325–331.
86. Sunardhiwidyaputra, S. and Vandamme, B. (1993) Immunohistochemical expression of tenascin in normal human salivary glands and in pleomorphic adenomas. *Pathol. Res. Practice*, **189**, 138–143.
87. Tanabe, K.K. and Saya, H. (1994) The CD44 adhesion molecule and metastasis. *Crit. Rev. Oncog.*, **5**, 201–212.
88. Tiitta, O., Wahlstrom, T., Virtanen, I. and Gould, V.E. (1993) Tenascin in inflammatory conditions and neoplasms of the urinary bladder *Virchows Archiv.*, **63**, 283–287.

89. Tiitta, O., Happonen, R.P., Virtanen, I. and Luomanen, M. (1994a) Distribution of tenascin in oral premalignant lesions and squamous cell carcinoma. *J. Oral Pathol. Med.*, **23**, 446–450.
90. Tiitta, O., Sipponen, P., Gould, V. and Virtanen, I. (1994b) Tenascin expression in inflammatory, dysplastic and neoplastic lesions of the human stomach. *Virchows Archiv.*, **425**, 369–374.
91. Truong, L.D., Pindur, J., Barrios, R., Dagati, V., Lechago, J., Suki, W. and Majesky, M. (1994) Tenascin is an important component of the glomerular extracellular matrix in normal and pathologic conditions. *Kidney Internat.*, **45**, 201–210.
92. Tucker, R.P., Hammerback, J.A. Jenrath, D.A., Mackie, E.J. and Xu, Y. (1993). Tenascin expression in the mouse: in situ localization and induction in vitro by bFGF. *J. Cell Sci.*, **104**, 69–73.
93. Tuominen, H. and Kallioinen, M. (1994) Increased tenascin expression in melanocytic tumors. *J. Cutaneous pathol.*, **21**, 424–429.
94. Vanduinen, C.M., Fleuren, G.J. and Bruijn, J.A. (1994) The extracellular matrix in pigmented skin lesions — An immunohistochemical study. *Histopathol.*, **24**, 33–40.
95. Vaughan, L., Weber, P., D'Alessandri, L., Zisch, A.H. and Winterhalter, K.H. (1994) Tenascin-contactin/Fll interactions: a clue for a developmental role? *Persp. Devel. Neurobiol.*, **2**, 43–52.
96. Vaupel, P., Kallinowski, F. and Okunieff, P. (1989) Blood flow, oxigen and nutrient supply, and methabolic microenvironment of human tumors: a review. *Cancer Res.*, **49**, 6449–6465.
97. Vollmer, G., Siegal, G.P., Chiquet-Ehrismann, R., Lightner, V. A., Arnholdt, H. and Knuppen, R. (1990) Tenascin expression in the human endometrium and in endometrial adenocarcinomas. *Lab. Invest.*, **62,** 725–730.
98. Vollmer, G., Lightner, V.A., Carter, C.A., Siegal, G.P., Erickson, H.P., Knuppen, R. and Kaufman, D.G. (1993) Localization of tenascin in uterine sarcomas and partially transformed endometrial stromal cells. *Pathobiol.*, **61**, 67–76.
99. Wood, E.J. Knaggs, H.E. and Cunliffe, W.J. (1991) Tenascin expression is induced during acne vulgaris. *Brit. J. Dermatol.*, **125**, 494.
100. Yamada, S., Ichida, T., Matsuda, Y., Miyazaki, Y., Hatano, T., Hata, K., Asakura, H., Hirota N., Geerts, A. and Wisse, E. (1992) Tenascin expression in human chronic liver disease and in hepatocellular carcinoma. *Liver*, **12**, 10–16.
101. Zagzag, D., Friedlander, D.R., Miller, D.C., Dosik, J., Cangiarella, J., Kostianovsky, M., Cohen, H., Grumet, M. and Greco, M.A. (1995) Tenascin expression in astrocytomas correlates with angiogenesis. *Cancer Res.*, **55**, 907–914.
102. Zardi, L., Carnemolla, B., Siri, A. Petersen, T.E., Paolella, G., Sebastio, G. and Baralle, F.E. (1987). Transformed human cells produce a new fibronectin isoform by preferential alternative splicing of a previously unobserved exon. *EMBO J.*, **6**, 2337–2342.
103. Zisch, A.H., D'Alessandri, L., Ranscht, B., Falchetto, R., Winterhalter, K.H. and Vaughan, L. (1992) Neuronal cell adhesion molecule contactin/Fll binds to tenascin via its immunoglobulin-domains. *J. Cell Biol.*, **119**, 203–213.

5. THROMBOSPONDINS

JACK LAWLER

Vascular Research Division, Department of Pathology, Brigham and Women's Hospital and Harvard Medical School, Boston, MA 02115

INTRODUCTION

The thrombospondins are a family of extracellular, calcium-binding proteins (for reviews see Adams *et al.*, 1995 and Lahav, 1993). Although five members of this gene family have been identified so far, the vast majority of functional characterization studies have used thrombospondin-1 from human blood platelets. Thrombospondin-1 has both adhesive and counteradhesive activities. In addition, it regulates cellular proliferation and migration. Through its various functions, thrombospondin-1 is reportedly involved in platelet aggregation, atherosclerosis, restenosis, inflammation, angiogenesis, wound healing, and tumor growth and metastasis.

In the vasculature, thrombospondin-1 supports the proliferation and migration of smooth muscle cells (Majack *et al.*, 1986 and 1988; Yabkowitz *et al.*, 1993). The thrombospondin-1 gene is rapidly up-regulated when smooth muscle cells are treated with platelet-derived growth factor (PDGF). Pretreatment of the cells with cycloheximide results in a further increase in the level of thrombospondin-1 mRNA. Thus, thrombospondin-1 behaves as an immediate early gene in response to PDGF. Secreted thrombospondin-1 serves as an autocrine factor for smooth muscle cells in that it acts synergistically with growth factors to stimulate progression through the cell cycle (Majack *et al.*, 1986).

In contrast to the growth promoting activity of thrombospondin-1 for smooth muscle cells is its inhibition of endothelial cell proliferation (Taraboletti *et al.*, 1990). This inhibition is thought to underlie the anti-angiogenic effects of thrombospondin-1. Angiogenesis *in vivo* is regulated by a balance between naturally occurring stimulators and inhibitors (Folkman, 1995). Thrombospondin-1 inhibits angiogenesis induced by basic fibroblast growth factor in a rat cornea model (Good *et al.*, 1990). The absence of thrombospondin-1 results in an increased angiogenic response during wound healing (Polverini *et al.*, 1995). In addition, the down-regulation of thrombospondin-1 in p53-deficient fibroblasts correlates with a switch to a proangiogenic phenotype (Dameron *et al.*, 1994). Since the growth of tumors is dependent upon blood flow, the decrease in thrombospondin-1 and the resulting shift in the balance of regulation of angiogenesis may represent a critical step in tumor progression.

Thrombospondin-1 is classified as a counteradhesive molecule because it inhibits the adhesion of endothelial cells to fibronectin and disrupts focal contacts that have been formed by endothelial cells (Lahav, 1988; Murphy-Ullrich and Hook, 1989). In contrast, it supports cellular attachment of several different cell types (see below). In general, the amount of thrombospondin-1 required to support cellular attachment is comparable to that required for the members of the adhesive glycoprotein family. Attachment of some cell types to thrombospondin-1 results in an organization of the actin cytoskeleton; however, focal contacts are rarely observed on thrombospondin-1-coated substrates (Adams and Lawler, 1994). In this chapter, I will review the structural and functional organization of the thrombospondins and summarize the properties of cell- surface receptors that interact with the thrombospondins.

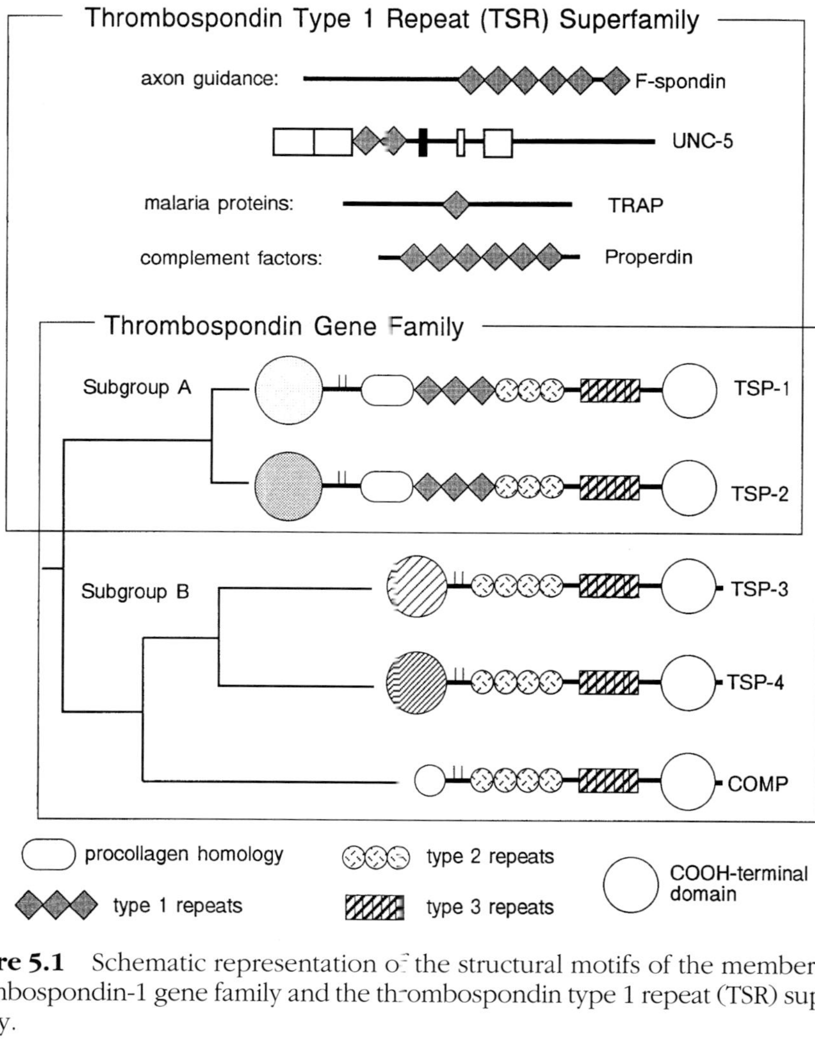

Figure 5.1 Schematic representation of the structural motifs of the members of the thrombospondin-1 gene family and the thrombospondin type 1 repeat (TSR) supergene family.

DOMAIN STRUCTURE OF THE THROMBOSPONDINS

The five known members of the thrombospondin gene family are designated thrombospondin-1, -2, -3, and -4 and cartilage oligomeric matrix protein (COMP) (Bornstein, 1992; Adams and Lawler, 1993b). The structural relationship among these proteins is shown in Figure 5.1. The thrombospondins can be divided into two subgroups on the basis of their domain structure. The subgroup A thrombospondins contain a region of homology with procollagen and the type 1 repeats. The subgroup B thrombospondins lack these two regions but have an additional type 2 repeat. In addition, COMP, which is a member of subgroup B, does not have the NH_2-terminal globular domain that is found in all of the other members of the thrombospondin gene family.

The NH_2-terminal domains of the thrombospondins have a lower level of sequence identity, than other regions of the molecule. These domains are similar to one another in size and in the position of two cysteine residues. The NH_2-terminal domains of thrombospondin-1 and -2 bind heparin with high affinity (Table 5.1) (O'Rourke *et al.*, 1992). Affinity co-electrophoresis

TABLE 5.1 Functional Domains of Thrombospondin-1

Domain	*Function*	*Receptors*
NH_2-terminal	Heparin binding Cell attachment, spreading and migration Disruption of focal contacts Regulation of proliferation Uptake and clearance Platelet aggregation	Heparin-sulfate proteoglycans Sulfatides
Procollagen homology	Anti-angiogenesis	
Type 1 repeats	Protein-protein interactions Heparin binding Cell attachment Neurite outgrowth TGF-β activation	CD36 50,000-dalton polypeptide Heparin-sulfate proteoglycans
Type 2 repeats	Calcium binding	
Type 3 repeats	Calcium binding Cell attachment Enzyme inhibition	β_3 integrins
COOH-terminal	Cell attachment and migration Platelet aggregation	80,000/105,000-dalton polypeptides 60,000-dalton polypeptides

and spin polarization measurements indicate that the dissociation constant for the binding of heparin to thrombospondin-1 is approximately 10 nM (San Antonio *et al.*, 1993; Slayter *et al.*, 1987). Two clusters of amino acids with positively charged side chains have been shown to mediate high-affinity binding of heparin to thrombospondin-1 (Table 5.2) (Lawler *et al.*, 1992).

TABLE 5.2 Active Sequences within Thrombospondin-1

Domain	*Sequence*	*Function*
NH_2-terminal	ELTGAARKGSGRRLVKGPD	Counteradhesion
	RKGSGRR, RQMKKTR	Heparin binding
Procollagen homology	NGVQYRN	Anti-angiogenesis
Type 1 repeats	CSVTCG	Cell binding
	WSXWSXW	Heparin binding
	RFK	TGF-β activation
	SPWDICSVTCGGGVQKRSK	Anti-angiogenesis
Type 2 repeats	$DVDEC(X)_6C(X)_8CENTDPGYNCLPC$	Calcium binding
Type 3 repeats	DXDXDXXXDXXD	Calcium binding
	RGD	Cell binding
	DNCPFHYNP, DNCQYVYNV	Enzyme inhibition
COOH-terminal	FYVVMWK, IRVVM	Cell binding

Mutation of these amino acids results in a decrease in the affinity of the protein for heparin. The NH_2-terminal domain also binds to sulfated glycolipids (Roberts *et al.*, 1985). Through these interactions, thrombospondin-1 can bind to cell-membrane proteoglycans and lipids. Specific interactions with the proteoglycans syndican and decorin have been demonstrated (Winnemoller *et al.*, 1992; Sun *et al.*, 1989). Murphy-Ullrich and Mosher (1987) have proposed that the interaction of thrombospondin-1 with cell membranes via its heparin-binding sites is involved in its uptake and degradation. The half-life of thrombospondin-1 in the extracellular environment is reportedly much shorter than that of fibronectin. The reason may be that thrombospondin-1 plays a regulatory role in cellular growth as opposed to a structural role in the extracellular matrix.

Thrombospondin-2, -3, and -4 have also been shown to bind heparin (O'Rourke *et al.*, 1992; Qabar *et al.*, 1994; Chen *et al.*, 1994; Lawler *et al.*, 1995). This activity involves the NH_2-terminal domain of thrombospondin-2. The high-affinity heparin-binding sites for thrombospondin-3 and -4 have not yet been localized. COMP reportedly is not retained by the heparin-Sepharose column during chromatography of EDTA extracts (DiCesare *et al.*, 1994). However, low-affinity heparin binding of some portions of thrombospondin-

1 appears to be calcium dependent (Lawler *et al.*, 1992). Thus, the possibility that COMP binds heparin cannot be ruled out.

The two interchain disulfide bonds and a region of homology with procollagen are immediately adjacent to the NH_2-terminal domain of thrombospondin-1 and -2 (Lawler and Hynes, 1986). This region of homology with procollagen is a structural motif that has recently been identified in other proteins (Francois *et al.*, 1994; Nelson *et al.*, 1994). In thrombospondin-1 and -2, this region appears to align the three subunits so that the interchain disulfide bonds are formed and the molecule is assembled into a trimer (Figure 5.2) (Prochownik *et al.*, 1989; Sottile *et al.*, 1991; Lawler *et al.*, 1992). Deletion

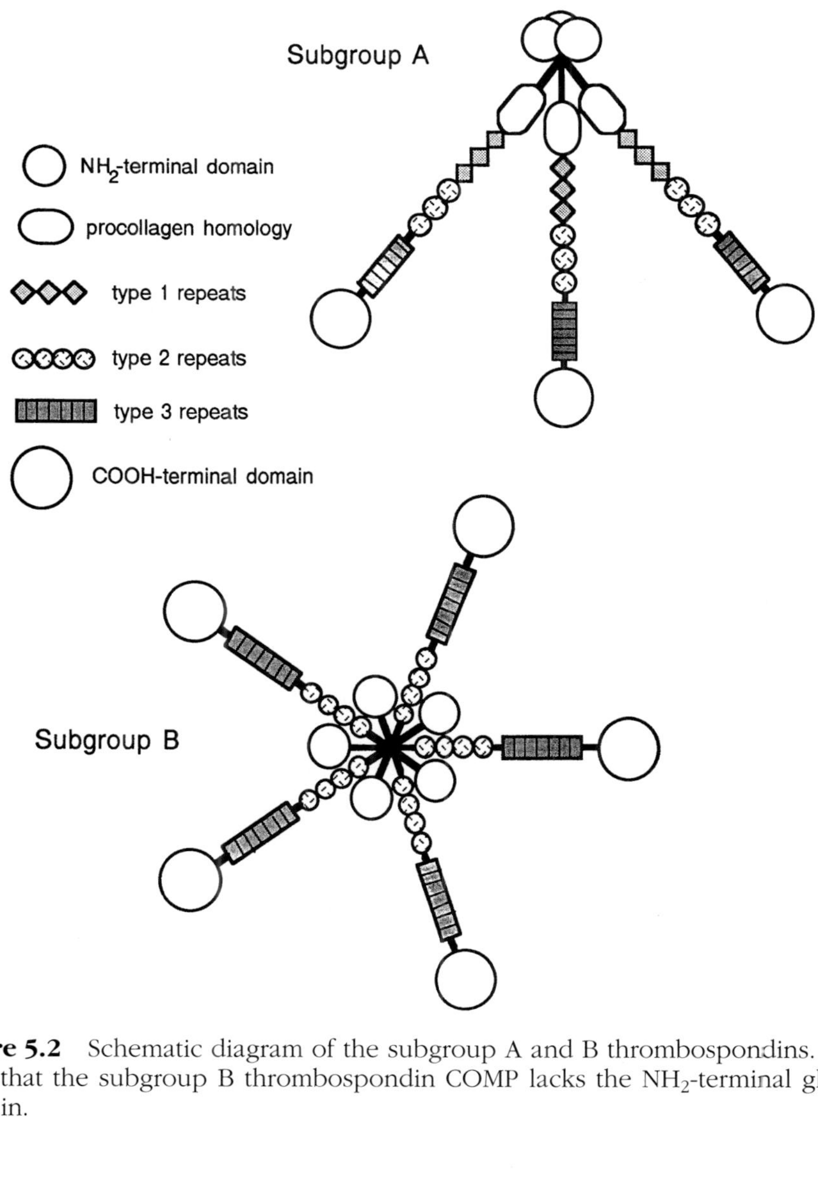

Figure 5.2 Schematic diagram of the subgroup A and B thrombospondins. Note that the subgroup B thrombospondin COMP lacks the NH_2-terminal globular domain.

of this region results in defective trimer assembly. The subgroup B thrombospondins contain two cysteine residues in a position similar to that of those forming the interchain disulfide bonds of the subgroup A thrombospondins; however, the region of homology with procollagen is lacking. The amino acid sequence immediately upstream of the two cysteine residues serves to align the subunits of the subgroup B thrombospondins (Efimov *et al.*, 1994). As a result of differences in the local structure around these cysteine residues, the subgroup B thrombospondins assemble into pentamers (Figure 5.2). In addition, this region reportedly binds type V collagen (Takaji *et al.*, 1993).

The subgroup A thrombospondins contain three copies of the type 1 repeat. A varied group of proteins comprise the thrombospondin type 1 repeat (TSR) supergene family (Figure 5.1). These include the complement factors C8, C9, and properdin as well as proteins that are produced by malaria-parasitized erythrocytes. Two proteins involved in axon guidance, UNC-5 and F-spondin, have recently been shown to be members of the TSR supergene family (Leung-Hagesteijn *et al.*, 1992; Klar *et al.*, 1992). The type 1 repeats of thrombospondin-1 have been shown to (1) contain a cellular adhesion sequence, (2) bind heparin and (3) bind and activate transforming growth factor-β (TGF-β). Several cell types express receptors that bind to the VTCG sequence of the type 1 repeats (Rich *et al.*, 1990; Prater *et al.*, 1991; Asch *et al.*, 1992; Li *et al.*, 1993; Tuszynski *et al.*, 1992). One of these proteins, CD36, also binds collagen and malaria-parasitized erythrocytes. The binding of CD36 to thrombospondin-1 and collagen is reportedly regulated by a novel mechanism that involves phosphorylation of an extracellular domain (Asch *et al.*, 1993). Whereas phosphorylated CD36 preferentially binds thrombospondin-1, dephosphorylated CD36 binds collagen. The identification of CD36 as a receptor for thrombospondin-1 is controversial. In support of this interaction, it has been shown that (1) antibodies to CD36 inhibit the thrombospondin-1-dependent adhesion of monocytes to platelets, (2) CD36 binds to thrombospondin-1 in solid-phase binding assays, (3) liposomes that contain CD36 bind to thrombospondin-1, and (4) antibodies to thrombospondin-1 immunoprecipitate a complex that contains CD36 as well as thrombospondin (Asch *et al.*, 1991; Stomski *et al.*, 1992; Legrand *et al.*, 1992; Greenwalt *et al.*, 1992). The data indicating that CD36 is not a receptor for thrombospondin-1 include the following: (1) COS cells transfected with CD36 do not bind thrombospondin-1. (2) Antibodies to CD36 do not inhibit the adhesion of platelets to solid-phase thrombospondin-1. (3) Nak^{a}-negative platelets lack CD36 and bind normally to thrombospondin-1 (Greenwalt *et al.*, 1992; Tandon *et al.*, 1991). Greenwalt *et al.* (1992) have proposed that CD36 is not the receptor for thrombospondin-1 but is in close proximity to or associated with the actual receptor. This proximity results in co-immunoprecipitation of CD36 with thrombospondin-1 bound to its receptor. In addition, antibodies to CD36 may sterically hinder access to the receptor.

Tuszynski and co-workers (1993) have recently identified a 50,000-dalton polypeptide on adenocarcinoma cells that binds to the CSVTCG sequence within the type 1 repeats. The binding of this protein to thrombospondin-1 is inhibited by EDTA and by heparin. A coordinate up-regulation of thrombospondin-1 and the 50,000-dalton polypeptide is evident in primary breast ductal carcinomas (Tuszynski and Nicosia, 1994). Capillary endothelium also expresses the 50,000-dalton polypeptide in the regions around the tumors.

The CSVTCG sequence is preceded by a sequence that contains highly conserved tryptophan residues (WSXWSXW). This sequence reportedly functions as a heparin-binding site (Guo *et al.*, 1992). The affinity for heparin is higher if additional upstream sequences are included. This site is involved in melanoma cell attachment to thrombospondin-1.

A sequence from the type 1 repeats is also involved in the activation of TGF-β by thrombospondin-1 (Murphy-Ullrich and Hook, 1989; Schultz-Cherry *et al.*, 1994). The sequence from the second type 1 repeat (amino acid residues 350-365) of thrombospondin-1 binds and activates TGF-β. Deletion of the first four amino acid residues renders the peptide incapable of activating TGF-β, although it still binds TGF-β (Schultz-Cherry *et al.*, 1995). Since comparable regions of mouse thrombospondin-2 do not activate TGF-β, and since the subgroup B thrombospondins lack the type 1 repeats, the ability to activate TGF-β appears to be specific for thrombospondin-1.

The subgroup A thrombospondins contain three copies of the epidermal growth factor (EGF)-like type 2 repeats, while the subgroup B thrombospondins contain four copies (Figure 5.1). These repeats are found in many extracellular matrix molecules and in the selectin family of cellular adhesion molecules (Engel, 1991). They are highly stable structural motifs that extend the polypeptide and confer resistance to proteases. In addition, all of the thrombospondins contain at least one type 2 repeat that conforms to a consensus site for calcium binding (Hanford *et al.*, 1991).

All of the thrombospondins are similar in the structure of the type 3 repeats and the COOH-terminal domain. The type 3 repeats are a contiguous set of calcium-binding sites. Each repeat contains a characteristic spacing of aspartic acid residues that positions oxygen atoms for calcium binding (Lawler and Hynes, 1986). Misenheimer and Mosher (1995) recently measured the binding of calcium to thrombospondin-1 by equilibrium dialysis. These authors found that a thrombospondin-1 molecule binds 35±3 calcium ions with an average dissociation constant of 50 μM at 4°C. This value is consistent with the value of 120 μM at 25°C that was determined in another study by titration of the circular dichroism spectrum (Lawler and Simons, 1982). Both studies revealed that the binding of calcium to thrombospondin-1 is cooperative. The subgroup A thrombospondins can bind 36 calcium ions while the subgroup B thrombospondins can bind 60 calcium ions because they are pentameric. The binding of large amounts of calcium by the thrombospondins indicates that one of the functions of these proteins may be to sequester calcium ions at the cell surface or in the extracellular matrix. Under appropriate circumstances, the calcium ions could be transferred to other calcium binding proteins to modulate their function. Calcium may also enter the cell and serve to enhance signal transduction.

Thrombospondin-1 also regulates fibrinolysis through its interactions with plasminogen, urokinase, and histidine-rich glycoprotein (Anonick *et al.*, 1993; Depoli *et al.*, 1989; Harpel *et al.*, 1990; Leung *et al.*, 1984; Mosher *et al.*, 1992; Silverstein *et al.*, 1984 and 1985). Solid-phase binding assays indicate that thrombospondin-1 binds fibrinogen, plasminogen, histidine-rich glycoprotein and both single and two-chain forms of urokinase. Thrombospondin forms a ternary complex with plasminogen and tissue plasminogen activator (Silverstein *et al.*, 1984). The formation of this complex results in more rapid plasmin generation. Thrombospondin-1 bound plasmin is protected from inhibition by α_2-plasmin inhibitor. Similarly, urokinase is protected from inhibition by plasminogen activator inhibitor type 1 when it is bound to thrombospondin-1. Whereas thrombospondin-1 increases the interval over which plasmin remains active by protecting it from inhibitors, plasmin is not proteolytically active while it is bound to thrombospondin-1. Thus, thrombospondin-1 can regulate plasmin generation and activity at specific sites on the cell surface and in the extracellular matrix. This regulation may be important during tumor cell metastasis, angiogenesis, leukocyte transmigration, and wound healing.

Thrombospondin-1 reportedly inhibits the activity of the neutrophil enzymes elastase and cathepsin G (Hogg, 1994). Thrombospondin-1 binding to elastase and cathepsin G occurred with dissociation constants of 17nM and 3.5nM, respectively. Each subunit of thrombospondin-1 bound one or two molecules of elestase or cathepsin G, respectively. The binding sites for these enzymes are located in the type 3 repeats (Table 5.2).

All of the thrombospondin-1 and -2 sequences contain an RGD cell- binding sequence within the type 3 repeats. The attachment of many cell types to thrombospondin-1-coated substrates is calcium dependent (Stomski *et al.*, 1992; Taraboletti *et al.*, 1990; Sun *et al.*, 1992; Lawler *et al.*, 1988). In addition, the attachment of fibroblasts to thrombospondin-2 is calcium-dependent (Chen *et al.*, 1994). The attachment of some cell types to thrombospondin-1 is inhibited by RGD-containing peptides (Tuszynski *et al.*, 1989; Asch *et al.*, 1991; Stomski *et al.*, 1992; Sun *et al.*, 1992; Adams and Lawler, 1993a; Lawler *et al.*, 1988). Moreover, attachment of cell types that are sensitive to RGD-containing peptides is inhibited by antibodies to the $\alpha_v\beta_3$ integrin. The RGD sequence is present in some of the subgroup B thrombospondins; however, it is not well conserved in different species. For example, human thrombospondin-4 contains an RGD sequence, but *Xenopus laevis* thrombospondin-4 does not.

An additional cell-binding site is found in the COOH-terminal domain of the thrombospondins (Kosfeld and Frazier, 1992 and 1993; Yabkowitz and Dixit, 1991). Antibodies to the COOH-terminal inhibit cellular attachment to thrombospondin-1 and platelet aggregation (Dixit *et al.*, 1985; Adams and Lawler, 1993a and 1994). A sequence found in all thrombospondins (IRVVM) has been identified as a binding site for melanoma cells (Kosfeld and Frazier, 1992 and 1993). A similar sequence (QRVVM) is found in some of the thrombospondins and is also functional as a cell-binding site. A 60,000-dalton protein on the surface of melanoma cells has been shown to act as a receptor for these sequences. In addition, Yabkowitz and Dixit (1991) have identified two polypeptides with molecular weights of 85,000 and 105,000 that function as a receptor for the COOH-terminal domain of thrombospondin-1.

For several cell types, it appears that the various receptor systems act concurrently (Asch *et al.*, 1991; Stomski *et al.*, 1992; Adams and Lawler, 1993a). Thus, G361 melanoma cell attachment to thrombospondin-1 can be inhibited with heparin or RGD-containing peptides. This implies that binding to an integrin and a proteoglycan are required simultaneously, since antagonists of either interaction are sufficient to block the interaction. The clustering of various cell surface binding molecules may be an integral part of the signalling events that may follow thrombospondin binding.

COUNTERADHESIVE PROPERTIES OF THROMBOSPONDIN-1

The counteradhesive activity of thrombospondin-1 has been demonstrated by treating fibronectin-coated substrates with thrombospondin-1 and by adding it to cells that are already attached and spread. Thrombospondin-1 supports attachment of endothelial cells; however, fewer cells are attached as compared to fibronectin-coated substrates and the cells do not spread (Lahav 1988; Lawler *et al.*, 1988; Taraboletti *et al.*, 1990). Endothelial cell attachment reportedly involves the binding of the RGD sequence to β_3 integrins (Lawler *et al.*, 1988; Taraboletti *et al.*, 1990). When fibronectin-coated substrates are treated with thrombospondin-1, the number of endothelial cells that attach is significantly inhibited (Lahav, 1988).

Like tenascin, the intact thrombospondin-1 molecule has the ability to disrupt focal contacts that are formed by endothelial cells on fibronectin-coated substrates (Murphy-Ullrich *et al.*, 1991 and 1993; Sage and Bornstein, 1991). The counteradhesive activity of thrombospondin-1 is neutralized by a monoclonal antibody to the NH_2-terminal heparin-binding domain (Murphy-Ullrich *et al.*, 1993). In addition, heparin neutralizes the ability of thrombospondin-1 to disrupt focal contacts. These observations prompted Murphy-Ullrich and co-workers (1993) to assay various peptides from the heparin-domain for the ability to disrupt focal contacts. A peptide (amino acids 17 to 35), designated hepI, reportedly displays counteradhesive activity, while other peptides from the heparin-binding domain are inactive. This peptide sequence is rich in amino acids that have positively charged side chains (arginine and lysine). These residues are essential components of the heparin-binding site of many proteins (for a review, see Cardin and Weintraub, 1989). Mutagenesis of the lysine and arginine residues in this sequence diminishes the ability of thrombospondin-1 to bind heparin (Lawler *et al.*, 1992). Modification of the lysine residues also renders the hepI peptide inactive (Murphy-Ullrich *et al.*, 1993). However, the capacity to disrupt focal contacts does not correlate with heparin-binding activity. Whereas scrambling of the amino acid sequence of the hepI peptide results in a loss of the ability to disrupt focal contacts, the scrambled peptide retains the ability to bind heparin. This observation indicates that the positively charged side chains must be positioned in a specific way for the peptide to express counteradhesive activity. In addition, it implies that a cell surface proteoglycan containing heparin side chains does not mediate the counteradhesive activity of thrombospondin-1. This conclusion is further supported by the ability of the hepI peptide to disrupt focal contacts formed by cells that have been treated with heparitinase (Murphy-Ullrich *et al.*, 1993).

The counteradhesive activity of thrombospondin-1 may be involved in the inhibition of endothelial cell mitogenesis and proliferation. Vogel and co-workers (1993) have shown that a recombinant form of the heparin-binding domain inhibits endothelial cell mitogenesis and proliferation. In addition, a peptide (amino acid residues 412 to 428) from the type 1 repeats also inhibits these endothelial cell functions (Vogel *et al.*, 1993). Two peptides (amino acid residues 424 to 442 and 481 to 491) from the second and third type 1 repeats inhibits angiogenesis in the rat cornea model and inhibits the migration of cultured endothelial cells (Tolsma *et al.*, 1993). These aspects of endothelial function are also inhibited by a peptide (amino acid residues 303 to 309) from the region of homology with procollagen (Tolsma *et al.*, 1993). These data indicate that multiple sites within the thrombospondin-1 molecule are involved in the modulation of endothelial cell phenotype.

CONCLUSION

Thrombospondin-1 has been shown to have both adhesive and counter-adhesive activities. The distinction between these two activities seems to depend largely on the type of cell under investigation. Each cell type expresses a different repertoire of thrombospondin-1 receptors on its surface. By interacting concurrently with several different membrane proteins, thrombospondin-1 directs the assembly of protein ensembles. These ensembles may include growth factors and cytokines. The composition of the ensembles may specifically signal phenotypic changes in the cells. In much the same way that complex ensembles of transcription factors are able precisely to regulate protein biosynthesis, cellular phenotype is determined by the integration of the information that is received from the extracellular protein ensembles on both a chemical and physical level (Hynes, 1992; Ingber, 1993).

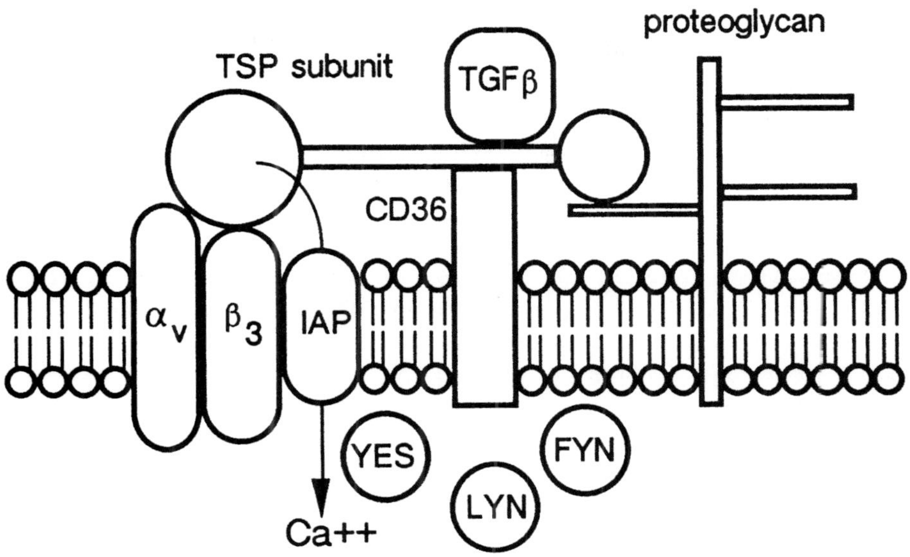

Figure 5.3 Schematic representation of protein ensembles on the cell surface.

Protein ensembles that are formed by thrombospondin-1 can include proteoglycans, CD36, $\alpha_v\beta_3$, and TGF-β (Figure 5.3). These proteins in turn can recruit other proteins to the complex. CD36 is reportedly associated with the Fyn, Lyn, and Yes protein-tyrosine kinases (Haung *et al.*, 1991). The $\alpha_v\beta_3$ integrin associates with membrane lipids and with membrane glycoproteins (Conforti *et al.*, 1990; Lindberg *et al.*, 1993). The integrin-associated protein (IAP) is a 50,000-dalton protein that copurifies with the $\alpha_v\beta_3$ integrin (Lindberg *et al.*, 1993). IAP reportedly regulates calcium entry into endothelial cells by functioning as a calcium channel (Schwartz *et al.*, 1993). Thrombospondin-1 may provide a localized source of calcium for entry into the cell via IAP. This calcium may facilitate growth factor stimulation or may stimulate a more migratory phenotype.

The multimeric nature of the thrombospondins may be essential for the complete assembly of membrane protein ensembles. Several point mutations and deletions of single amino acids within the type 3 repeats of COMP have been shown to cause pseudoachondroplasia (Hecht *et al.*, 1995). This abnormal phenotype has an autosomal dominant pattern of inheritance; thus, the effected individuals presumably have one normal and one wild type allele. Since COMP is a pentamer, the molecules will be composed of 0–5 normal or abnormal subunits. The presence of a variable number of abnormal subunits appears to be sufficient to perturb function. These results indicate that the multimeric structure of COMP is important for the normal function of the protein.

An understanding of the molecular mechanisms that underlie counteradhesive activity will enhance our ability to incorporate this activity into a general scheme for regulation of cellular phenotype. Focal contacts formed by endothelial cells can be disrupted with a short peptide from the thrombospondin-1 sequence. Whereas this sequence mediates heparin binding to thrombospondin-1, the ability of the peptide to bind heparin does not correlate with the ability to disrupt focal contacts. A second peptide, designated hepII (amino acids 74–95), does not disrupt focal contacts (Murphy-Ullrich *et al.*, 1993). Mutagenesis of the basic amino acid residues within this region or within the sequence included in the inhibiting peptide (hepI) inhibits heparin binding to intact thrombospondin-1 (Lawler *et al.*, 1992). Thus, only one of the two sites involved in the binding of intact thrombospondin-1 to heparin is involved in the disruption of focal contacts. The heparin-binding domain of thrombospondin-1 also binds to sulfatides (Roberts *et al.*, 1985). The potential role of sulfatides in the counteradhesive activity of thrombospondin-1 remains to be determined. The molecular mechanisms underlying counteradhesiveness are at the heart of the role of the extracellular matrix in determining cellular phenotype.

ACKNOWLEDGEMENTS

I would like to thank Joan Smith for help in preparing the manuscript. This work was supported by the National Institutes of Health research grants HL28749 and HL49081 from the National Heart, Lung and Blood Institute.

REFERENCES

1. Adams, J.C. and Lawler, J. (1993a) Diverse mechanisms for cell attachment to platelet thrombospondin. *J. Cell Sci.* **104**, 1061–1071.
2. Adams, J.C. and Lawler, J. (1993b) The thrombospondin family. *Curr. Biol.* **3**, 188–190.
3. Adams, J.C. and Lawler, J. (1994) Cell type specific adhesive interactions of skeletal myoblasts with thrombospondin-1. *Mol. Biol. Cell* **5**, 423–437.
4. Adams, J.C., Tucker, R. and Lawler, J. (1995) The thrombospondin gene family. R.G. Landes Co., Austin.
5. Anonick, P.K., Yoo, J.K., Webb, D.J. and Gonias, S.L. (1993) Characterization of the antiplasmin activity of human thrombospondin-1.
6. Asch, A.S., Liu, I., Briccetti, F.M., Barnwell, J.W., Kwakye-Berko, F., Dokun, A., Goldberg, J. and Pernambuco, M. (1993) Analysis of CD36 binding domains: Ligand specificity controlled by dephosphorylation of an ecto-domain. *Science* **262**, 1436–1440.
7. Asch, A.S., Silbiger, S., Heimer, E. and Nachman, R.L. (1992) Thrombospondin sequence motif (CSVTCG) is responsible for CD36 binding. *Biochem. Biophys. Res. Comm.* **182**, 1208–1217.

8. Asch, A.S., Tepler, J., Silbiger, S. and Nachman, R.L. (1991) Cellular attachment to thrombospondin: Cooperative interactions between receptor systems. *J. Biol. Chem.* **266**, 1740–1745.
9. Bornstein, P. (1992) Thrombospondins: Structure and regulation of expression. *FASEB J.* **6**, 3290–3299.
10. Cardin, A.D. and Weintraub, H.J.R. (1989) Molecular modeling of protein-glycosaminoglycan interactions. *Arteriosclerosis* **9**, 21–31.
11. Chen, H., Sottile, J., O'Rourke, K.M., Dixit, V.M. and Mosher, D.F. (1994) Properties of recombinant mouse thrombospondin 2 expressed in *Spodoptera* cells. *J. Biol. Chem.* **269**, 32226–32232.
12. Conforti, G., Zanetti, A., Pasquali-Ronchetti, I., Quaglino, Jr., D., Neyroz, P. and Dejane, E. (1990) Modulation of vitronectin receptor binding by membrane lipid composition. *J. Biol. Chem.* **265**, 4011–4019.
13. Dameron, K.M., Volpert, O.V., Tainsky, M.A. and Bouck, N. (1994) Control of angiogenesis in fibroblasts by p53 regulation of thrombospondin-1. *Science* **265**, 1582–1584.
14. DePoli, P., Bacon-Baguley, T., Kendra-Franczak, S., Cederholm, M.T. and Walz, D.A. (1989) Thrombospondin interaction with plasminogen. Evidence for binding to a specific region of the kringle structure of plasminogen. *Blood* **73**, 976–982.
15. DiCesare, P.E., Morgelin, M., Mann, K. and Paulsson, M. (1994) Cartilage oligomeric matrix protein and thrombospondin-1: Purification from articular cartilage, electron microscopic structure, and chondrocyte binding. *Eur. J. Biochem.* **223**, 927–937.
16. Dixit, V.M., Haverstick, D.M., O'Rourke, K,M., Hennessy, S.W., Grant, G.A., Santoro, S.A. and Frazier, W.A. (1985) A monoclonal antibody against human thrombospondin inhibits platelet aggregation. *Proc. Natl. Acad. Sci. USA.* **82**, 3472–3476.
17. Efimov, V.P., Lustig, A. and Engel, J. (1994) The thrombospondin-like chains of cartilage oligomeric matrix protein are assembled by a five-stranded α-helical bundle between residues 20 and 83. *FEBS Letters* **341**, 54–58.
18. Engel, J. (1991) Domains in proteins and proteoglycans of the extracellular matrix with functions in assembly and cellular activities. *Int. J. Biol. Macro. Mol.* **13**, 147–151.
19. Folkman, J. (1995) Angiogenesis in cancer, vascular, rheumatoid and other disease. *Nature Medicine* **1**, 27–31.
20. Francois, V., Solloway, M., O'Neill, J.W., Emery, J. and Bier, E. (1994) Dorsal-ventral patterning of the *Drosophila* embryo depends on a putative negative growth factor encoded by the short gastrulation gene. *Genes Dev.* **8**, 2602–2616.
21. Good D.J., Poverini, P.J., Rastinejad, F., LeBean, M.M., Lemons, R.S., Frazier, W.A. and Bouck, N.P. (1990) A tumor-suppressor-dependent inhibitor of angiogenesis is immunologically and functionally indistinguishable from a fragment of thrombospondin. *Proc. Natl. Acad. Sci. USA* **87**, 6624–6628.
22. Greenwalt, D.E., Libsky, R.H., Ockenhouse, C.F., Ikeda, H., Tandon, N.M. and Jamieson, G.A. (1992) Membrane glycoprotein CD36: A review of its roles in adherence, signal transduction and transfusion medicine. *Blood* **80**, 1105–1115.
23. Guo, N.-H., Krutzsch, H.C., Negre, E., Vogel, T., Blake, D.A. and Roberts, D.D. (1992) Heparin- and sulfatide-binding peptides from the type 1 repeats of human thrombospondin promote melanoma cell adhesion. *Proc. Natl. Acad. Sci. USA* **89**, 3040–3044.

24. Hanford, P.A., Mayhew, M. and Baron, M. (1991) Key residues involved in calcium-binding motifs and EGF-like domains. *Nature* **351**, 164–167.
25. Harpel, P.C., Silverstein, R.L., Pannell, R., Gurewich, V. and Nachman, R.L. (1990) Thrombospondin forms complexes with single-chain and two-chain forms of urokinase. *J. Biol. Chem.* **265**, 11289–11294.
26. Haung, M.-M., Bolen, J.B., Barnwell, J.W., Shattil, S.J. and Brugge, J.S. (1991) Membrane glycoprotein IV (CD36) is physically associated with Fyn, Lyn and Yes protein-tyrosine kinases in human blood platelets. *Proc. Natl. Acad. Sci. USA* **88**, 7844–7848.
27. Hecht, J.T., Nelson, L.D., Crowder, E., Wang, Y., Elder, F.F.B., Harrison, W.R., Francomano, C.A., Prange, C.K., Lennon, G.G., Deere, M. and Lawler, J. (1995) Mutations in exon 17B of cartilage oligomeric matrix protein (COMP) cause pseudoachondroplasia. *Nature Genetics* **10**, 325–329.
28. Hogg, P.J. (1994) Thrombospondin 1 as an enzyme inhibitor. *Thromb. Haemos.* **72**, 787–792.
29. Hynes, R.O. (1992) Integrins: versatility, modulation and signaling in cell adhesion. *Cell* **69**, 11–25.
30. Ingber, D. (1991) Integrins as mechanochemical transducers. *Cur. Opin. Cell Biol.* **3**, 841–848.
31. Klar, A., Baldassare, M. and Jessell, T.M. (1992) F-spondin: a gene expressed at high levels in the floor plate encodes a secreted protein that promotes neural cell adhesion and neurite extension. *Cell* **69**, 95–110.
32. Kosfeld, M.D. and Frazier, W.A. (1992) Identification of active peptide sequences in the carboxyl-terminal domain of human thrombospondin-1. *J. Biol. Chem.* **267**, 16230–16236.
33. Kosfeld, M.D. and Frazier, W.A. (1993) Identification of a new cell adhesion motif in two homologous peptides from the COOH-terminal cell binding domain of human thrombospondin. *J. Biol. Chem.* **268**, 8808–8814.
34. Lahav, J. (1988) Thrombospondin inhibits adhesion of endothelial cells. *Exp. Cell Res.* **177**, 199–204.
35. Lahav, J. (1993) Thrombospondin. CRC Press, Boca Raton.
36. Lawler, J., Ferro, P. and Duquette, M. (1992) Expression and mutagenesis of thrombospondin. Biochemistry **31**, 1173–1180.
37. Lawler, J. and Hynes, R.O. (1986) The structure of human thrombospondin, an adhesive glycoprotein with multiple calcium binding sites and homologies with several different proteins. *J. Cell Biol.* **103**, 1635–1648.
38. Lawler, J., McHenry, K., Duquette, M. and Derick, L. (1995) Characterization of human thrombospondin-4. *J. Biol. Chem.* **270**, 2809–2814.
39. Lawler, J. and Simons, E. (1982) Cooperative binding of calcium to thrombospondin. The effect of calcium on the circular dichroism and limited tryptic digestion of thrombospondin. *J. Biol. Chem.* **258**, 12098–12101.
40. Lawler J., Weinstein, R. and Hynes, R.O. (1988) Cell attachment to thrombospondin: The role of Arg-Gly-Asp, calcium and integrin receptors. *J. Cell Biol.* **107**, 2351–2361.
41. Legrand, C., Thibert, V., Dubernard, V., Begault, B. and Lawler, J. (1992) Molecular requirements for the interaction of thrombospondin with thrombin-activated human platelets: Modulation of platelet aggregation. *Blood* **79**, 1995–2003.

42. Leung, L.L.K., Nachman, R.L. and Harpel, P.C. (1984) Complex formation of platelet thrombospondin with histidine-rich glycoprotein. *J. Clin. Invest.* **73**, 5–12.
43. Leung-Hagesteijn, C., Spense, A.M., Stern, B.D., Zhou, Y., Su, M.-W., Hedgecock, E.M. and Culotti, J.G. (1992) UNC-5, a transmembrane protein with immunoglobulin and thrombospondin type 1 domains, guides cell and pioneer axon migrations in C. elegans. *Cell* **71**, 289–299.
44. Li, W.-X., Howard, R.J. and Leung, L.K. (1993) Identification of SVTCG in thrombospondin as the conformation-dependent, high affinity binding site for its receptor, CD36. *J. Biol. Chem.* **268**, 16179–16184.
45. Lindberg, G.P., Greshaw, H.D., Schartz, M.A. and Brown, E.J. (1993) Molecular cloning of integrin-associated protein: An immunoglobulin family member with multiple membrane spanning domains implicated in $\alpha_v\beta_3$-dependent ligand binding. *J. Cell Biol.* **123**, 485–486.
46. Majack, R.A., Cook, S.K. and Bornstein, P. (1986) Regulation of smooth muscle cell growth by components of the extracellular matrix: an autocrine role for thrombospondin. *Proc. Natl. Acad. Sci. USA* **83**, 9050–9054.
47. Majack, R.A., Goodman, I.V. and Dixit, V.M. (1988) Cell surface thrombospondin is functionally essential for vascular smooth muscle cell proliferation. *J. Cell Biol.* **106**, 415–422.
48. Misenheimer, T.M. and Mosher, D.F. (1995) Calcium ion binding to thrombospondin-1. *J. Biol. Chem.* **270**, 1729–1733.
49. Mosher, D.F., Misenheimer, T.M., Stenflo, J. and Hogg, P.J. (1992) Modulation of fibrinolysis by thrombospondin. *Ann. N.Y. Acad. Sci.* **667**, 64–69.
50. Murphy-Ullrich, J.E. and Hook, M. (1989) Thrombospondin modulates focal adhesions in endothelial cells. *J. Cell Biol.* **109**, 1309–1319.
51. Murphy-Ullrich, J.E. and Mosher, D.F. (1987) Interactions of thrombospondin with endothelial cells: receptor-mediated binding and degradation. *J. Cell Biol.* **105**, 1603–1611.
52. Murphy-Ullrich, J.E., Gurusiddappa, S., Frazier, W.A. and Hook, M. (1993) Heparin-binding peptides from thrombospondins 1 and 2 contain focal adhesion-labilizing activity. *J. Biol. Chem.* **268**, 26784–26789.
53. Murphy-Ullrich, J.E., Lightner, V.A., Aukhil, I., Yan, Y.Z., Erickson, H.P. and Hook, M. (1991) Focal adhesion integrity is down regulated by the alternatively spliced domain of human tenascin. *J. Cell Biol.* **115**, 1127–1136.
54. Nelson, R.E., Fessler, L.I., Takagi, Y., Blomberg, B., Keene, D.R. Olson, P.F., Parker, C.G. and Fessler, J.H. (1994) Peroxidasin: a novel enzyme-matrix protein of *Drosophila* development. *Eur. J. Biochem.* **13**, 3438–3447.
55. O'Rourke, K.M., Laherty, C.D. and Dixit, V.M. (1992) Thrombospondin 1 and thrombospondin 2 are expressed as both homo-and heterotrimers. *J. Biol. Chem.* **265**, 24921–24924.
56. Polverini, P.J., DiPietro, L.A., Dixit, V.M., Hynes, R.O. and Lawler, J. (1995) Thrombospondin-1 knockout mice showed delayed organization and prolonged neovascularization of skin wounds. *FASEB J.* **9(3)** Part I, A272.
57. Prater, C.A., Plotkin, J., Jaye, D. and Frazier, W.A. (1991) The properdin-like type 1 repeats of human thrombospondin contain a cell attachment site. *J. Cell Biol.* **112**, 1031–1040.

58. Prochownik, E.V., O'Rourke, K. and Dixit, V.M. (1989) Expression and analysis of COOH-terminal deletions of the human thrombospondin molecule. *J. Cell Biol.* **109**, 843–852.
59. Qabar, A.N., Lin, Z., Wolf, F.W., O'Shea, K.S., Lawler, J. and Dixit, V.M. (1994) Thrombospondin-3 is a developmentally regulated heparin binding protein. *J. Biol. Chem.* **269**, 1262–1269.
60. Rich, K.A., George, F.W., IV, Law, J.L. and Martin, W.J. (1990) Cell-adhesive motif in region II of malarial circumsporozoite protein. *Science* **249**, 1574–1577.
61. Roberts, D.D., Haverstick, D.M., Dixit, V.M., Frazier, W.A., Santoro, S.A. and Ginsburg, V. (1985) The platelet glycoprotein thrombospondin binds specifically to sulfated glycolipids. *J. Biol. Chem.* **260**, 9405–9411.
62. Sage, E.H. and Bornstein, P. (1991) Extracellular proteins that modulate cell-matrix interactions. *J. Biol. Chem* **266**, 14831–14834.
63. San Antonio, J.D., Slover, J., Lawler, J., Karnovsky, M.J., and Lander, A.D. (1993) Specificity in the interactions of extracellular matrix protein with subpopulations of the glycosaminoglycan heparin. *Biochemistry* **32**, 4746–4755.
64. Schultz-Cherry, S., Chen, H., Mosher, D.F., Misenheimer, T.M., Krutzsch, H.C., Roberts, D.D., Murphy-Ullrich, J.E. (1995) Regulation of transforming growth factor-β activation by discrete sequences of thrombospondin-1. *J. Biol. Chem.* **270**, 7304–7310.
65. Schultz-Cherry, S., Lawler, J. and Murphy-Ullrich, J.E. (1994) The type 1 repeats of thrombospondin-1 activate TGF-β. *J. Biol. Chem.* **269**, 26783–26788.
66. Schwartz, M.A., Brown, E.J. and Fazeli, B. (1993) A 50-kd integrin-associated protein is required for integrin-regulated calcium entry in endothelial cells. *J. Biol. Chem.* **268**, 19931–19934.
67. Silverstein, R.L., Leung, L.L.K., Harpel, P.C. and Nachman, R.L. (1984) Complex formation of platelet thrombospondin with plasminogen: Modulation of activation by tissue factor. *J. Clin. Invest.* **74**, 1625–1633.
68. Silverstein, R.L., Leung, L.L.K., Harpel, P.C. and Nachman, R.L. (1985) Platelet thrombospondin forms a trimolecular complex with plasminogen and histidine-rich glycoprotein. *J. Clin. Invest.* **75**, 2065–2073.
69. Slayter, H.S., Karp, G., Miller, B.E. and Rosenberg, R.D. (1987) Binding properties of human thrombospondin: Interaction with mucopolysaccharides. *Semin. Thrombosis and Hemostasis* **13**, 369–377.
70. Sottile, J., Seleque, J. and Mosher, D.F. (1991) Synthesis of truncated amino terminal trimers of thrombospondin. *Biochemistry* **30**, 6556–6562.
71. Stomski, F.C., Gani, J.S., Bates, R.C. and Burns, G.F. (1992) Adhesion to thrombospondin by human embryonic fibroblasts is mediated by multiple receptors and includes a role for glycoprotein 88 (CD36). *Exp. Cell Res.* **198**, 85–92.
72. Sun, X., Mosher, D.F. and Rapraeger, A. (1989) Heparin sulfate-mediated binding of epithelial cell surface proteoglycan to thrombospondin. *J. Biol. Chem.* **264**, 2885–2889.
73. Sun, X., Skorstengaard, K. and Mosher, D.F. (1992) Disulfides modulate RGD-inhibitable adhesive activity of thrombospondin. *J. Cell Biol.* **118**, 693–701.
74. Takaji, J., Fujisawa, T., Usui, T., Aoyama, T. and Saito, Y. (1993) A single chain 19-kDa fragment from bovine thrombospondin binds to type V collagen and heparin. *J. Biol. Chem.* **268**, 15544–15549.

75. Tandon, N.N., Ockenhouse, C.F., Greco, N.J. and Jamieson, G.A. (1991) Adhesive functions of platelets lacking GPIV (CD36). *Blood* **78**, 2809–2813.
76. Taraboletti, G., Roberts, D., Liotta, L.A. and Giavazzi, R. (1990) Platelet thrombospondin modulates endothelial cell adhesion, motility and growth: a potential angiogenesis regulatory factor. *J. Cell Biol.* **111**, 765–772.
77. Tolsma, S.S., Volpert, O.V., Good, D.J., Frazier, W.A., Polverini, P.J., and Bouck, N. (1993) Peptides derived from two separate domains of the matrix protein thrombospondin-1 have anti-angiogenic activity. *J. Cell Biol.* **122**, 497–511.
78. Tuszynski, G.P. and Nicosia, R.F. (1994) Location of thrombospondin and its cysteine-serine-valine-threonine-cysteine-glycine-specific receptor in human breast carcinoma. *Lab. Invest.* **70**, 228–233.
79. Tuszynski, G.P., Karczewski, J., Smith, L., Murphy, A., Rothman, V.L. and Knudsen, K.A. (1989) The GPIIb-IIIa-like complex may function as a human melanoma cell adhesion receptor for thrombospondin. *Exp. Cell. Res.* **182**, 473–481.
80. Tuszynski, G.P., Rothman, V.L., Deutch, A.H., Hamilton, B.K. and Eyal, J. (1992) Biological activities of peptides and peptide analogues derived from common sequences present in thrombospondin, properdin and malaria proteins. *J. Cell Biol.* **116**, 209–217.
81. Tuszynski, G.P., Rothman, V.L., Papole, M., Hamilton, B.K. and Eyal, J. (1993) Identification and characterization of a tumor cell receptor for CSVTCG, a thrombospondin adhesive domain. *J. Cell Biol.* **120**, 513–521.
82. Vogel, T., Guo, N-K, Krutzsch, H.C., Blake, D.A., Hartman, J., Mendelovitz, S., Panel, A. and Roberts, D.D. (1993) Modulation of endothelial cell proliferation, adhesion and motility by recombinant heparin-binding domain and synthetic peptides form the type 1 repeats of thrombospondin. *J. Cell Biochem.* **53**, 74–84.
83. Winnemoller, M., Schon, P., Vischer, P. and Kresse, H. (1992) Interactions between thrombospondin and the small proteoglycan decorin: interference with cell attachment. *European J. Cell Biol.* **59**, 47–55.
84. Yabkowitz, R. and Dixit, V. (1991) Human carcinoma cells bind thrombospondin through a M_r80,000/105,000 receptor. *Cancer Res.* **51**, 3648–3656.
85. Yabkowitz, R., Mansfield, P.J., Ryan, V.S. and Suchard, S.J., (1993) Thrombospondin mediates migration and potentiates platelet-derived growth factor-dependent migration of calf pulmonary artery smooth muscle cells. *J. Cell. Physiol.* **157**, 24–32.

6. SPARC AS A COUNTERADHESIVE PROTEIN: STRUCTURAL AND FUNCTIONAL CORRELATES

KOUROS MOTAMED, JAMES A. BASSUK and E. HELENE SAGE

Department of Biological Structure, University of Washington, Seattle, WA 98195

INTRODUCTION

SPARC is an extracellular matrix-associated glycoprotein with counteradhesive properties similar to those of tenascin and thrombospondin. Although these three proteins are structurally unrelated, they appear to serve dynamic roles in embryogenesis, morphogenesis, and tissue remodeling. It has been proposed that SPARC, and in some cases thrombospondin and tenascin, mediates transient binding to specific components of the extracellular matrix but does not constitute a structural matrix component in normal adult vertebrates (Lane and Sage, 1994). SPARC (secreted protein, acidic and rich in cysteine), also termed osteonectin, 43 K protein, and basement membrane protein BM-40, is a serum albumin- and collagen-binding glycoprotein. The protein was first isolated by Termine *et al.* (1981) as a major noncollagenous constituent of bovine and human bone and dentin. Subsequently, SPARC was found to be identical to a 43 K protein synthesized *in vitro* as part of a "culture shock" response to conditions promoting proliferation and migration of endothelial cells (Sage *et al.*, 1984) and fibroblasts (Otsuka *et al.*, 1984). Later, Mason *et al.* (1986a) showed that SPARC is a major secreted product of differentiated mouse embryo parietal endoderm cells that are specialized for the synthesis of Reichert's membrane. cDNA sequence comparisons have shown that osteonectin, 43K, BM-40, and SPARC are identical proteins or interspecies homologs (Mann *et al.*, 1987; Mason *et al.*, 1986b; Lankat-Buttgereit *et al.*, 1988; Swaroop *et al.*, 1988).

STRUCTURE AND REGULATION OF THE SPARC GENE

Chromosome mapping in both vertebrates (Mason *et al.*, 1986b; Swaroop *et al.*, 1988) and nematode worms (Schwarzbauer and Spencer 1993) has identified SPARC as a single-copy gene. In these studies, genes for several cytokines, growth factors, and growth factor receptors have been shown to map close to the SPARC locus. To date, however, no developmental abnormalities have been shown to be the direct result of mutations within the SPARC locus.

Nucleotide sequence comparison of molecularly cloned SPARC genes revealed high degrees of sequence identity among bovine, mouse, and human SPARC (Villarreal *et al.*, 1989). The percent nucleotide identity between human and mouse SPARC is 75% for the 5´ noncoding region, 90% for the coding region, and 61% for the 3´ noncoding region. A comparison of nucleotide sequences from human and bovine SPARC revealed 87% identity for the 5´ noncoding, 93% for the coding, and 76% for the 3´ noncoding regions. Bovine, murine, and human SPARC genes are greater than 20 kb in size, and all consist of 10 exons separated by 9 introns (McVey *et al.*, 1988; Villarreal *et al.*, 1989). A more recently cloned nematode SPARC exhibits clear structural differences, in that it contains only six exons, lacks exons 1, 3, and 10, and has undergone fusion of exons 6 and 7 (Schwarzbauer and Spencer, 1993). Despite the absence of intron 1, the remaining intron-exon boundaries are apparently conserved, in comparison to the mammalian counterparts.

Sequencing analysis of bovine, mouse, and human genes has revealed an absence of canonical TATA and CAAT box sequences, which are thought to be involved in the correct positioning of RNA polymerase (McVey *et al.*, 1988). Nevertheless, only a single major, and one minor, transcriptional start site for SPARC has been identified in a variety of cells by both RNase protection and primer extension studies (McVey *et al.*, 1988). The 5´ flanking region and exon 1 of the mouse and bovine SPARC genes contain one or more potential cis-acting regulatory elements: e.g., tandem GC-rich motifs as potential Sp-1 transcription factor-binding sites, transcription factor AP-2 binding sites, and response elements specific for cAMP, retinoic acid, heat shock, and growth hormone (McVey *et al.*, 1988; Young *et al.*, 1989). Exon 1 is untranslated and contains endonuclease S1-sensitive, repetitive purine-rich "GCA" boxes that have been implicated in cell-specific regulation of the SPARC gene (Young *et al.*, 1989). This repeating DNA sequence, clustered approximately 20 base pairs upstream of the translational start site, exhibits an unusually high degree of conservation among species. Young *et al.* (1989) recently described an apparently novel 40 kDa cell type-specific GGA-binding factor in bovine bone cells that, by DNaseI footprinting analysis, interacted with the 3´ end of the SPARC promoter.

Northern blot analysis of both total and poly(A)$^+$ RNA from murine, bovine, and human tissues and cells has revealed a major 2.2 kb SPARC transcript, and a less abundant 3.0 kb transcript (Mason *et al.*, 1986b; Swaroop *et al.*, 1988). Interestingly, human basal and differentiated keratinocytes, as well as breast adenocarcinoma cells, exhibited an abundant expression of a 3.0 kb transcript but no detectable 2.2 kb transcript (Howe *et al.*, 1990). To date, no splice variants of SPARC have been reported, and the two alternative mRNAs are thought to result from differential polyadenylation in cells from different animal species (Swaroop *et al.*, 1988). Shorter mRNA sizes of 1.6 kb and 1.1 kb have been reported in frogs and nematodes, respectively (Damjanovski *et al.*, 1992; Schwarzbauer and Spencer, 1993). High levels of SPARC RNA are usually found in tissues undergoing synthesis and remodeling of extracellular matrix (Holland *et al.*, 1987; Sage *et al.*, 1989). Conversely, more limited transcription of the SPARC gene has been reported in cells transformed with Simian virus-40, v-abl, and c-src (Mason *et al.*, 1986a).

Several cultured cell types, such as fetal bovine bone and skin cells, undifferentiated murine F9 embryonal carcinoma cells, and differentiated parietal endoderm cells, have been used successfully to determine the genetic regulatory elements responsible for the complex differential expression pattern of the SPARC gene. *cis*-acting elements regulating tissue-specific patterns of expression during differentiation of bovine bone and skin cells, as well as mouse parietal endoderm cells, have been delineated. A summary of the findings of these transient transfection studies is shown in Table 6.1.

Table 6.1 cis-acting Regulatory Elements in the SPARC Promoter[a]

Source	*Promoter region*	*FBBC*	*FBSC*	*PYS/F9*
Bovine	–287 to +11	2	1.5	
	–504 to +11	50	6.0	
	–927 to +11	NC	NC	
Murine	–78 to +11			NC
	–169 to +11			3.2
	–356 to +11			6.3
	–465 to +11			1.5

[a]Cultures of FBBC (fetal bovine bone cells), FBSC (fetal bovine skin cells), Fg, and PYS (murine parietal yolk sac endoderm cells) were transiently transfected with various segments of the SPARC promoter ligated to a CAT reporter gene. Promoter activity is shown as relative CAT activity over vector alone (for FBBS and FBSC) (Dominguez *et al.*, 1991), or as a ratio of relative CAT activity in PYS cells compared to that of F9 embryonal carcinoma cells prior to differentiation (Nomura *et al.*, 1989). NC, no change.

Morphogens, cytokines, and hormones have also been reported to induce changes in SPARC expression during development (Nomura *et al.*, 1989; Ng *et al.*, 1989). Direct evidence for induction of transcription was provided by studies employing progesterone and dexamethasone in rat preosteoblasts (Ng *et al.*, 1989) and human breast epithelial cells (Nomura *et al.*, 1989). Indirect stimulation of SPARC expression by growth factors such as transforming growth factor-β, platelet-derived growth factor, insulin-like growth factor-1, and colony stimulating factor-1 has also been reported (Chandrasekhar *et al.*, 1994). The cytokine interleukin-1, implicated in enhanced matrix destruction and in blocking matrix repair in degenerative diseases, also acts as an inhibitor of SPARC synthesis (Chandrasekhar *et al.*, 1994). Our present level of knowledge of the regulatory factors directing SPARC expression cannot fully account for the significant changes in the level of the protein observed during development and remodeling in adult tissues. A more detailed understanding of the regulation of the SPARC gene is therefore requisite.

EXPRESSION PATTERN OF SPARC

SPARC is a 43 kDa Ca^{+2}-binding glycoprotein that has been conserved during the evolution of nematodes, amphibians, birds, and mammals (primates, rodents, and artiodactyls). Amphibian and nematode SPARC proteins are 80% and 38% identical to mammalian SPARC, respectively. Since SPARC is a secreted glycoprotein, delineation of its exact spatial and temporal expression during development requires localization of both mRNA and protein by *in situ* hybridization and by immunohistochemistry. In Xenopus, SPARC transcripts and protein are expressed first during the early development of notochord and somites, and are maximal in the intersomitic clefts as long as the myotomes remain electrically coupled by gap junctions (Ringuette *et al.*, 1991). Several developmental defects were observed when anti-SPARC antibodies were injected into developing embryos, or as a result of ectopic overexpression of SPARC. In mice, increased expression of SPARC in the gut, bone, and gonads of adults was interpreted as an indication that SPARC is probably associated with tissue remodeling, cell movement, and/or proliferation in some organs (Sage *et al.*, 1989). In *C. elegans*, overexpression of SPARC in muscle cells resulted in an "uncoordinated" *(Unc)* phenotype, in which adults moved abnormally or became paralyzed (Schwarzbauer and Spencer, 1993). These results predict a role for SPARC in the modulation of cell-matrix interactions, cell attachment, and/or synthesis and turnover of extracellular matrix. High levels of SPARC transcripts and protein are also found in developing bones and teeth of bovine, murine, and human embryos (Mundlos *et al.*, 1992). Elevated production of SPARC is also detected when attachment-dependent endothelial cells, fibroblasts, and glial cells are placed in culture, a phenomenon referred to as "culture shock" (Sage *et al.*, 1986). Increased synthesis of SPARC protein has also been associated with heat shock (Neri *et al.*, 1992), cellular aging *in vitro* (Murano *et al.*, 1992), and wound repair (Reed *et al.*, 1993). These observations indicate that SPARC might affect the expresssion of matrix components which in turn regulate cellular interactions with the extracellular matrix.

STRUCTURAL DOMAINS OF SPARC

Vertebrate SPARC cDNAs encode proteins of 298–304 amino acids, with at least 4 distinct domains based on the predicted secondary structure (Engel *et al.*, 1987) (Figure 6.1). A highly acidic domain I (amino acids 3–51) follows a 17 amino acid signal sequence and is encoded by exons 3 and 4. This

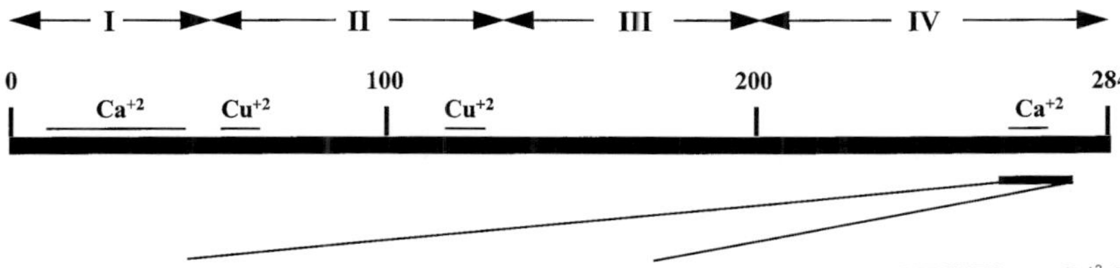

MUTATION	SEQUENCE (254 … 273; S--S between C255 and C271; boxed D L D N D K Y I A L E E; W*)	ACTIVITY	Ca+2-BINDING
4.0 wt	TTRFFE T C D L D N D K Y I A L E E W* A G C F G IKEQDINKDLVI-COOH	+	+
4.2 wt	- - - - - - - - - - - - - - - - - - - -	+	+
4.2 K	- - - - K - - - - - - - - - - - - - - -	-	-
4.2 ND	- - - - N - - - - D - - - - - - - - - -	+	+
4.2 AA	- A - - - - - - - - - - - - - - - A - -	-	-
D257K	.... K	ND	+
E268K	.... E	ND	-
C256/272-L	.... L . . . . L	ND	-

Figure 6.1 Amino acid residues of the EF-hand motif of SPARC critical for Ca^{+2}-binding and biological activity. Shown are the sequence and numbering of the secreted form of murine SPARC. The 4 domains of SPARC are indicated by Roman numerals. The regions that bind Ca^{+2} or Cu^{+2} are designated by bars (—). Peptides 4.0, 4.2 wild type (wt), 4.2K, 4.2 ND, and 4.2AA are synthetic peptides of mouse SPARC (Sage *et al.*, 1995). Peptides that contain amino acid substitutions are 4.2K, 4.2ND, and 4.2AA (e.g., 4.2K is identical to 4.2wt, except that D258 has been replaced with a K residue). D257K, E268K, and C256/272-L are point substitutions in human rSPARC (Pottgiesser *et al.*, 1994). Note that the numbering of D257K, E268K, and C256/272-L corresponds to the sequence of human SPARC, which is offset by +2 from the mouse SPARC numbering indicated. Biological activity is defined as the inhibition of DNA synthesis by endothelial cells (Sage *et al.*, 1995). Positive activity is denoted as "+". No or diminished activity is denoted as "-". Ca^{+2}-binding is defined according to the quenching of the UV fluorescence spectra of the peptide (Sage *et al.*, 1995) or rSPARC (Pottgiesser *et al.*, 1994). The Trp residue responsible for this fluorescence is designated (W*).

domain binds 5–8 Ca^{+2} with low affinity. Domain II (amino acids 52–132) is cysteine-rich and is encoded by exons 5 and 6. It contains sequences that are homologous to a repeated domain of follistatin, an inhibitor of the transforming growth factor-β-like cytokines activin and inhibin (Lane and Sage, 1994). Domain II also contains two Cu^{+2} -binding sites and the sequence GHK that has been implicated in the regulation of cell proliferation (Funk and Sage, 1991; Funk and Sage, 1993). Domain III (amino acids 133–227), encoded by exons 6 and 7, contains an endogenous protease-sensitive site and is composed of a series of α-helical segments. Domain IV (amino acids 228–285) comprises an EF-hand loop encoded by exon 9 that binds a single Ca^{+2} with high affinity. The EF-hand has been shown to mediate binding to collagen (Lane and Sage, 1990; Mayer *et al.*, 1991) and endothelial cells (Yost and Sage, 1993).

A FAMILY OF SPARC-LIKE GENES

Recently, cDNAs for two SPARC-related genes, SC1 (Johnston *et al.*, 1990) and QR1 (Guermah *et al.*, 1991), were isolated from a rat brain library and from embryonic quail retina, respectively. Despite encoding translation products approximately twice the size of SPARC, both SC1 and QR1 exhibit a C-terminal sequence of approximately 200 amino acids that is over 60% identical to mouse SPARC. The conservation of the SPARC-like EF-hand region implies that both proteins might bind Ca^{+2}. Moreover, it appears that the follistatin-like module of SPARC domain II, implicated in binding to cytokines, is retained in both SC1 and QR1. Both SPARC-like proteins appear to be correlated with cell growth and differentiation in the central nervous system. Based on immunological data, two other genes that share structural characteristics with SPARC have been identified: testican, a multidomain testicular proteoglycan (Alliel *et al.*, 1993) and gp60/albondin, an albumin-binding protein on the endothelial cell surface (Schnitzer and Oh, 1992). A graphical representation of the SPARC family of proteins is shown in Figure 6.2.

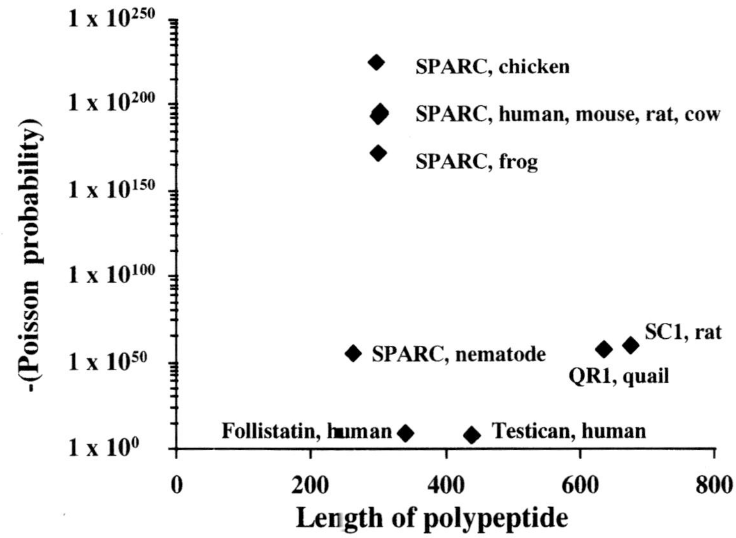

Figure 6.2 Analysis of SPARC and SPARC-like proteins. The amino acid sequence of chicken SPARC was used to query the non-redundant protein database (PDB + SwissProt + SPupdate + PIR + GenPept + GPupdate) according to the BLASTP computer program (Altschul *et al.*, 1990). Plotted on the ordinate is the Poisson probability that a score as high as the score shown would be observed by chance if a search of a polypeptide sequence of equal length were performed. Accession numbers are as follows: chicken SPARC (S39539), human SPARC (P09486), mouse SPARC (P19883), cow SPARC (P13213), quail QR1 (P23499), rat SC1 (P24054), human follistatin (P10669), and human testican (S33293). The sequence of rat SPARC is derived from unpublished data from our laboratory.

PROPERTIES OF SPARC

Specific Interaction with Ca^{+2}

The first observation of a specific, high-affinity interaction with Ca^{+2} was reported for SPARC prepared from EDTA extracts of adult bovine bone (Romberg *et al.*, 1985). The intrinsic ultraviolet fluorescence of the protein was partially quenched when micromolar concentrations of Ca^{+2} were added, a result indicative of a high-affinity interaction (K_d = 0.3 uM) with the metal. From the molecular cloning and sequencing of SPARC cDNA (Mason *et al.*, 1986b), this high-affinity site was assigned to a C-terminal EF-hand motif (Engel *et al.*, 1987). A second EF-hand-like motif in domain III has been proposed (Bolander *et al.*, 1988), although its detection is algorithm-dependent (Bassuk *et al.*, 1993). Sequence analysis of SPARC cDNA also indicated that several Ca^{+2} ions could bind to the negatively-charged domain I (Engel *et al.*, 1987), and subsequent calculations have assigned the Kd as ~10 mM (Maurer *et al.*, 1992). Fusion proteins between β-galactosidase and either domain I or domain IV of nematode SPARC were shown to bind Ca^{+2}; thus, each end of the SPARC protein interacts with Ca^{+2} (Schwarzbauer and Spencer, 1993). The anti-spreading activity of domain I (Lane and Sage, 1990) can be linked to the property of this region to bind Ca^{+2}. Since peptides from domain I have also been immunogenic, the epitopes are thought to be exposed in the native protein (Lane and Sage, 1990).

Recombinant SPARC Proteins and Their Properties

Recombinant SPARC (rSPARC) has been expressed in kidney 293 cells (Pottgiesser *et al.*, 1994), yeast (Yost *et al.*, 1994), and *E. coli* (Bassuk *et al.*, 1995). Since the isolated protein from each system is a monomer as revealed by non-reducing sodium dodecyl sulfate — polyacrylamide gel electrophoresis (SDS-PAGE), the polypeptide appears to be correctly folded. The bacterial and yeast proteins were biologically active, and both the bacterial and mammalian proteins displayed Ca^{+2}-dependent conformations (Table 6.2). rSPARC can therefore be expressed in heterologous systems and its biological activity linked to conformations that are Ca^{+2}-dependent. Like native SPARC, the conformation of rSPARC has been studied by intrinsic fluorescence emission and circular dichroism (CD) spectroscopies (Pottgiesser *et al.*, 1994; Bassuk *et al.*, 1995).

The deletion of the C-terminal EF-hand or its single disulfide bond decreased production and secretion of rSPARC from 293 cells, a result indicating that the C-terminal region of rSPARC is required for correct folding (Pottgiesser *et al.*, 1994). Point mutations made within, or adjacent to, the "loop" of the helix-loop-helix EF-hand (Figure 6.1) have linked this motif to the existence of one high-affinity Ca^{+2}-binding site. In addition, these mutations have confirmed the original suggestion by Engel *et al.* (1987) that a single disulfide bond serves to stabilize the motif. The EF-hand of SPARC is unusual, in that the

Table 6.2 Recombinant SPARC Proteins and Their Properties

Protein Source	*Expression Plasmid & promoter*	*Expression Host*	*Biological Activity*	*Conformational Activity*
Human Umbilical Vein Endothelium[a]	•plasmid pET22b •T7 RNA pol	*E. coli*	•receptor binding •inhibition of cell spreading •inhibition of DNA synthesis	•monomer by SDS-PAGE under nonreducing conditions •Ca^{+2}-dependent CD spectra •Ca^{+2}-dependent UV fluorescence
Human Placenta[b]	•plasmid pCMV •cytomegalovirus	human embryonal kidney 293 cell line	not determined	•monomer by SDS-PAGE under nonreducing conditions •Ca^{+2}-dependent CD spectra •Ca^{+2}-dependent UV fluorescence •Ca^{+2}-dependent collagen IV binding
Mouse Parietal Endoderm[c]	•plasmid pRPOT with BAR-SPARC fusion; •triosephosphate isomerase	*S. cerevisiae*	•receptor binding •inhibition of cell spreading	•monomer by SDS-PAGE under non-reducing conditions

[a](Bassuk *et al.*, 1995), [b](Nischt *et al.*, 1991; Pottgiesser *et al.*, 1994), [c](Yost *et al.*, 1994)

majority of EF-hands are found in cytosolic proteins and are clustered in groups of two or more. That a second EF-hand might exist in domain III of SPARC prompted Pottgiesser *et al.* (1994) to mutate the position X residue of this sequence (D175 to K); however, no significant effect on Ca^{+2} affinity was observed. Resolution of the issue of a second EF-hand must await the crystallization and subsequent determination of the 3-dimensional structure of SPARC.

Cellular proliferation assays with synthetic peptides containing mutated sequences of the Ca^{+2}-binding region of the EF-hand are consonant with the above studies in the linking of biological activity to occupation of the high-affinity site by Ca^{+2}, and to an intact disulfide bond (Figure 6.1; Sage *et al.*, 1995). When peptide 4.2 wt (Figure 6.1) was added to newly-plated endothelial cells or fibroblasts, cell spreading was inhibited in a concentration-dependent manner (Lane and Sage, 1990). The same peptide was found to be a potent inhibitor of endothelial cell DNA synthesis *via* a cooperative action with a peptide from domain II (peptide 2.1), itself a known inhibitor of cell proliferation (Sage *et al.*, 1995). Intrinsic fluorescence and cell proliferation studies with the wild-type and mutated SPARC peptides showed that both the conformation of peptide 4.2 and the Ca^{+2}-coordinating Asp at position 258 were necessary for the inhibition of DNA synthesis in cultured endothelial cells (Figures 6.1 and Figure 6.3) (Sage *et al.*, 1995).

Interaction with Ca^{+2} also increased the mean residue ellipticity of native SPARC (Engel *et al.*, 1987; Maurer *et al.*, 1992) and of recombinant SPARC in circular dichroism spectroscopic assays (Pottgiesser *et al.*, 1994; Bassuk *et al.*, 1995). Since increases in the mean residue ellipticity of proteins are positively correlated with increases in the amount of α-helicity (Chen *et al.*, 1972), we have concluded that the biologically active form of SPARC exhibits a conformation that is both rich in α-helical structure and is Ca^{+2}-dependent. Preparations of biologically active rSPARC from bacteria display this Ca^{+2}-dependent change in mean residue ellipticity (Bassuk *et al.*, 1995). Bacterial rSPARC binds in a specific fashion to smooth muscle cells (Funk *et al.*, 1995) and endothelial cells (Bassuk *et al.*, 1995). The change in the circular dichroic spectrum of mammalian rSPARC has also been associated with domain III, as large deletions of this domain significantly reduce the affinity of the protein for Ca^{+2} (Pottgiesser *et al.*, 1994). In addition to sequences that stimulate the production of matrix metalloproteinases (Tremble *et al.*, 1993) and bind collagen (Pottgiesser *et al.*, 1994), a new function for domain III might therefore be the transduction of the Ca^{+2} signal into a conformational change that results in a new biological activity.

Another issue relevant to rSPARC proteins is the functionality of posttranslationally-attached carbohydrate. Three different N-linked glycosylation sites have been reported for the seven SPARC and two SPARC-like proteins (Figure 6.2). All contain the N-linked glycosylation signal N-K/Q-T. Chick, human, and cow contain an additional signal (N-N-S/T), and nematode SPARC has

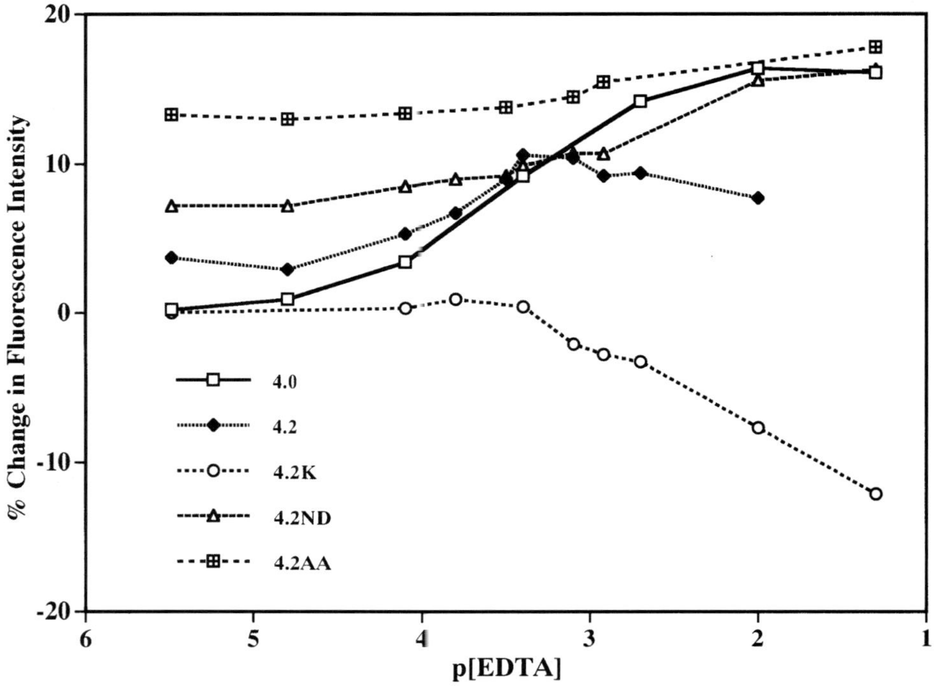

Figure 6.3 Altered intrinsic fluorescence of SPARC peptides from the C-terminal EF-hand region upon removal of bound Ca^{+2}. SPARC peptides were dissolved at a concentration of 0.8 – 1.4 mM into 0.01 M Tris-HCl/0.15 M NaCl/0.05 mM $CaCl_2$ (pH 9.0). After extensive dialysis against this buffer previously adsorbed on Chelex™ resin, samples were diluted to 20–30 mM in the same buffer that contained varying concentrations of EDTA. Shown are plots of percent change in intrinsic fluorescence at 354 nm as a function of pEDTA. Reproduced with permission of Wiley-Liss ©1995.

the sequence N-I-S. Are these signals required for biological activity? We know that human SPARC expressed and purified from bacteria lacks carbohydrate but nevertheless exhibits activity in three types of assays (Bassuk *et al.*, 1995). Moreover, a mutant mouse rSPARC, expressed in yeast and lacking the known N-linked carbohydrate attachment site, was counteradhesive (Yost *et al.*, 1994) and reduced focal contacts in bovine aortic endothelial cells (Murphy-Ullrich *et al.*, 1995). It appears, for these activities at least, that the carbohydrate is not a critical feature. An alternative explanation for the existence of carbohydrate on SPARC is the promotion of folding of the nascent polypeptide within the endoplasmic reticulum. The chaperones Bip and calnexin, the latter known to bind selectively to carbohydrates within glycoproteins, could bind SPARC to facilitate its proper folding and disulfide bond formation (Kim and Arvan, 1995; Helenius, 1994). The oligosaccharide might play a key role in the rapid and efficient secretion process associated with the posttranslational processing of SPARC. Since the endoplasmic reticulum is rich in stored Ca^{+2}, the form of SPARC that is allowed to leave this compartment is likely to be bound to Ca^{+2} and would therefore be secreted with a maximal content of α-helix.

FUNCTIONS OF SPARC

Cell shape and adhesion

SPARC and peptides from the Ca^{+2}-binding region of SPARC elicited a rounded morphology in endothelial cells, fibroblasts, and smooth muscle cells that was correlated with an RGD-independent inhibition of cellular spreading (Sage *et al.*, 1989; Sage and Bornstein, 1991; Murphy-Ullrich *et al.*, 1995). These effects of SPARC and its peptides were specific, concentration-dependent, and non-toxic; moreover, the inhibition of cell spreading did not involve sequestration of extracellular Ca^{+2} and Mg^{+2} (Sage, 1992). Supporting evidence for the role of SPARC in the regulation of cellular morphology was provided by studies of stably-transfected murine F9 embryonal carcinoma cell lines overexpressing sense or antisense SPARC cDNA (Everitt and Sage, 1992a). Cells that overexpressed SPARC were aggregated and rounded, whereas underexpressing lines spread to an even greater extent than that of the control cells transformed with vector alone. When these cell lines were cultured in Ca^{+2}-deficient media, SPARC-overexpressing cells attached and spread, whereas underexpressing lines attached but remained rounded (Everitt and Sage, 1992b). The observed enhanced spreading in the SPARC-overexpressing cells supports the notion that the native conformation of SPARC, and therefore its anti-spreading property, was compromised in a Ca^{+2}-deficient environment. Moreover, it was shown that in the presence Ca^{+2}-chelators, overexpression of SPARC mediated enhanced attachment and spreading of F9 cells under Ca^{+2}-deficient conditions. It is therefore possible that the adhesive functions of SPARC are dependent in part on the local concentration of extracellular Ca^{+2}.

Decreased requirements for Ca^{+2} have been reported to be associated with cellular transformation and metastasis (Means and Rasmussen, 1988). Interestingly, increased expression of SPARC has also been found in several invasive tumors *in vivo* (Mann *et al.*, 1987; Wewer *et al.*, 1988; Columbo *et al.*, 1991; Porter *et al.*, 1995). These observations suggest that SPARC could serve potentially as a promoter of transient cellular adhesion in a Ca^{+2}-deficient extracellular environment, a characteristic consistent with the proposed function of SPARC as a modulator of cell-cell and cell-matrix interactions in response to tissue remodeling or injury (Lane and Sage, 1994).

Barrier function and permeability

Selective transport of circulating macromolecules in and out of the extracellular environment is regulated by the vascular endothelium (Pober, 1988; Goldblum *et al.*, 1994). With cultured pulmonary vascular endothelial cells as a model system, it has been shown that SPARC, like tumor necrosis factor-α and bacterial endotoxin, induces changes in cell shape that are temporally coincident with actin reorganization, formation of intercellular gaps, and barrier dysfunction (Figure 6.4; Goldblum *et al.*, 1993a, 1993b, 1994). SPARC induced the depolymerization and redistribution of F-actin to the endothelial cell periphery. This cytoskeletal rearrangement appears to be a prerequisite to increases in endothelial cell permeability, since prior stabilization of actin with phallicidin

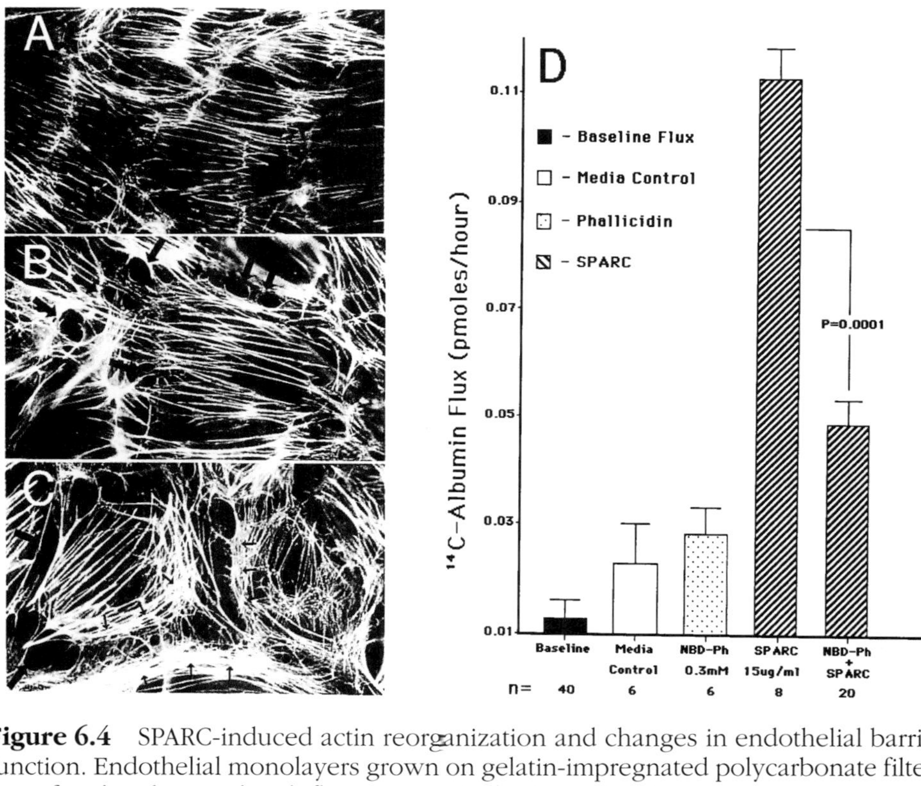

Figure 6.4 SPARC-induced actin reorganization and changes in endothelial barrier function. Endothelial monolayers grown on gelatin-impregnated polycarbonate filters were fixed and stained with fluorescein-phalloidin in the absence (A) or presence of SPARC at 15 μg/ml for a minimum of 2 h (B and C). Exposure of cells to SPARC resulted in the formation of intercellular gaps (closed arrows in B), and in a redistribution of circumferential F-actin (small closed arrows in C). Pretreatment of endothelial monolayers with NBD-phallicidin (NBD-Ph), which stabilizes the actin cytoskeleton, prior to and throughout a 6 h exposure to SPARC, protected endothelial monolayers against the SPARC-mediated increases in the transendothelial flux of ^{14}C-BSA. Reproduced with permission of Harwood Academic Press © 1994.

protected against changes in barrier function (Figure 6.4; Goldblum *et al.*, 1994). Exposure of endothelial cells to exogenous SPARC in the absence of serum increased the flux of labeled bovine serum albumin across postconfluent monolayers in a concentration-dependent manner (Goldblum *et al.*, 1994). This increase in cell permeability was blocked by anti-SPARC antibodies but not by inhibition of protein synthesis, and could not be ascribed to cell injury or loss of endothelial cell viability. These results are consistent with findings that SPARC diminished the number of focal adhesions in cultured bovine aortic endothelial cells in a concentration-dependent manner, with a corresponding redistribution of vinculin and actin-containing stress fibers (Murphy-Ullrich *et al.*, 1995). Moreover, it was shown that two SPARC peptides derived from domains II and IV, and known to compete for the binding of SPARC to endothelial cells, mediated focal adhesion disassembly, presumably through interaction with a cell-surface receptor(s) (Murphy-Ullrich *et al.*, 1995).

A more detailed understanding of the structure-function and signaling relationships between SPARC and changes in vascular endothelial barrier function is lacking. It would also be of great importance to determine whether SPARC mediates physiological and/or pathological regulation of barrier function *in vivo*.

SUMMARY AND CONCLUSIONS

In this short review we have focused on the structural features of SPARC most closely related to its counteradhesive functional properties. Highly relevant characteristics include a) relatively low-affinity Ca^{+2}-binding sites within the N-terminal domain (I), which might be sensitive to local, variable concentrations of extracellular Ca^{+2}, and b) an EF-hand in the C-terminal domain (IV) that binds a single Ca^{+2} with higher affinity and is stabilized by a disulfide bond. Although we have not identified the molecular pathways by which SPARC diminishes cellular adhesion, we can state that the mechanism is not one by which the protein acts as a simple chelator of divalent cations, although Ca^{+2} in particular is critical for the conformation and activity of SPARC. It is also apparent that a major target for SPARC is the focal adhesion complex on endothelial cells (and probably other cells) in culture. Following an interaction of SPARC with one or more components of this adhesive complex, with extracellular macromolecules associated with the complex, or potentially with its own receptor, SPARC effects a significant rearrangement of cytoskeleton, identified experimentally as the breakdown of actin-containing stress fibers and a redistribution of actin toward the cell periphery. A potentially important consequence of the cytoskeletal alteration induced by SPARC is the introduction of intercellular gaps and, in the case of the endothelium, an increase in transendothelial transport of macromolecules and hence in the permeability of this intrinsic physiological barrier.

Other functions of SPARC are most likely related to its counteradhesive function but have not been discussed in this review: a) direct inhibition of the cell cycle of endothelial cells, smooth muscle cells, and fibroblasts in the mid-to-late G1 phase, b) indirect inhibition of cell growth by an interaction with the B-chain of platelet-derived growth factor, c) inhibition of the mitogenic and chemotactic effect of basic fibroblast growth factor on endothelial cells, d) interaction with extracellular components including several collagen types and thrombospondin, and e) regulation of the production of thrombospondin, fibronectin, and plasminogen activator inhibitor-1 by endothelial cells. Future studies *in vivo* will clarify the relationship of these properties to the effects of SPARC on cell shape and cell adhesion, *e.g.*, in wound healing, angiogenesis, or cancer biology.

BIBLIOGRAPHY

1. Alliel, P.M., Perin, J.P., Jolles, P. and Bonnet, F.J. (1993) Testican, a multidomain testicular proteoglycan resembling modulators of cell social behaviour. *Eur. J. Biochem.* **21**, 347–350.
2. Altschul, S.F., Warren, G., Miller, W., Myers, E.W. and Lipman, D.J. (1990) Basic local alignment search tool. *J. Mol. Biol.* **215**, 403–410.
3. Bassuk, J.A., Iruela-Arispe, M.L., Lane, T.F., Benson, J.M., Berg, R.A. and Sage, E.H. (1993) Molecular analysis of chicken embryo SPARC (osteonectin). *Eur. J. Biochem.* **218**, 117–127.
4. Bassuk, J.A., Baneyx, F., Vernon, R.B., Funk, S.E. and Sage, E.H. (1995) Expression of biologically active human SPARC in Escherichia coli. *Arch. Biochem. Biophys.* **submitted.**
5. Bolander, M.E., Young, M.F., Fisher, L.W., Yamada, Y. and Termine, J.D. (1988) Osteonectin cDNA sequence reveals potential binding regions for calcium and hydroxyapatite and shows homologies with both a basement membrane protein (SPARC) and a serine proteinase inhibitor (ovomucoid). *Proc. Natl. Acad. Sci. U. S. A.* **85**, 2919–2923.
6. Chandrasekhar, S., Harvey, A.K., Johnson, M.G. and Becker, G.W. (1994) Osteonectin/SPARC is a product of articular chondrocytes/cartilage and is regulated by cytokines and growth factors. *Biochim. Biophys. Acta* **1221**, 7–14.
7. Chen, Y.H., Yang, Y.T., and Martinez, H.M. (1972) Determination of the secondary structure of proteins by circular dichroism and optical rotary dispersion. *Biochemistry* **11**, 4120–4131.
8. Columbo, M.P., Ferrari, G., Geraldina, B., Galasso, D., Howe, C.C. and Parmiani, G.P. (1991) Down-regulation of SPARC/osteonectin/BM-40 expression in methylcholanthrene-induced fibrosarcomas and in Kirsten-MSV transformed fibroblasts. *Eur. J. Cancer* **27**, 58–62.
9. Damjanovski, S., Liu, F. and Ringuette, M. (1992) Molecular analysis of Xenopus laevis SPARC. A highly conserved calcium-binding extracellular matrix protein. *Biochem. J.* **281**, 513–517.

10. Dominguez, P., Ibaraki, K., Robey, P.G., Hefferan, T.E., Termine, J.D. and Young, M.F. (1991) Expression of the osteonectin gene potentially controlled by multiple cis- and trans-acting factors in cultured bone cells. *J. Bone Miner. Res.* **6**, 1127–1136.
12. Engel, J., Taylor, W., Paulsson, M., Sage, H. and Hogan, B. (1987) Calcium-binding domains and calcium-induced transition in SPARC (osteonectin/BM-40), an extracellular glycoprotein expressed in mineralized bone and nonmineralized tissues. *Biochemistry* **26**, 6958–6965.
13. Everitt, E.A. and Sage, E.H. (1992a) Expression of SPARC is correlated with altered morphologies in transfected F9 embryonal carcinoma cells. *Exp. Cell Res.* **199**, 134–146.
14. Everitt, E.A. and Sage, E.H. (1992b) Overexpression of SPARC in stably transfected F9 cells mediates attachment and spreading in Ca^{+2}-deficient medium. *Biochem. Cell Biol.* **70**, 1368–1379.
15. Funk, S.E., Raines, E., Bassuk, J., Ross, R. and Sage, E.H. (1995) Interaction of SPARC with human smooth muscle cells. Abrogation of mitogen-stimulated proliferation. *J. Cell. Biochem.* **submitted.**
16. Funk, S.E. and Sage, E.H. (1991) The Ca^{-2}-binding glycoprotein SPARC modulates cell cycle progression in bovine aortic endothelial cells. *Proc. Natl. Acad. Sci. U.S.A.* **88**, 2648–2652.
17. Funk, S.E. and Sage, E.H. (1993) Differential effects of SPARC and cationic SPARC peptides on DNA synthesis by endothelial cells and fibroblasts. *J. Cell Physiol.* **154**, 53–63.
18. Goldblum, S.E., Ding, X., Brann, T.W. and Campbell-Washington, J. (1993a) Bacterial lipopolysaccharide induces actin reorganization, intercellular gap formation, and endothelial barrier dysfunction in pulmonary vascular endothelial cells: concurrent F-actin depolymerization and new actin synthesis. *J. Cell Physiol.* **157**, 13–23.
19. Goldblum, S.E., Ding, X. and Campbell-Washington, J. (1993b) TNF-alpha induces endothelial cell F-actin depolymerization, new actin synthesis, and barrier dysfunction. *Am. J. Physiol.* **264**, C894–C905.
20. Goldblum, S.E., Ding, X., Funk, S.E. and Sage, E.H. (1994) SPARC (secreted protein acidic and rich in cysteine) regulates endothelial cell shape and barrier function. *Proc. Natl. Acad. Sci. U.S.A.* **91**, 3448–3452.
21. Guermah, M., Crisanti, P., Laugier, D., Dezelee, P., Bidou, L., Pessac, B. and Calothy, G. (1991) Transcription of a quail gene expressed in embryonic retinal cells is shut off sharply at hatching. *Proc. Natl. Acad. Sci. U.S.A.* **88**, 4503–4507.
22. Helenius, A. (1994) How N-linked oligosaccharides affect glycoprotein folding in the endoplasmic reticulum. *Mol. Biol. Cell* **5**, 253–265.
23. Holland, P., Harper, S., McVey, J. and Hogan, B.L.M. (1987) In vivo expression of mRNA for the Ca^{+2}-binding protein SPARC (osteonectin) revealed by in situ hybridization. *J. Cell Biol.* **105**, 473–482.
24. Howe, C.C., Kath, R., Mancianti, M.L., Herlyn, M., Mueller, S. and Cristofalo, V. (1990) Expression and structure of human SPARC transcripts: SPARC mRNA is expressed by human cells involved in extracellular matrix production and some of these cells show an unusual expression pattern. *Exp. Cell Res.* **188**, 185–191.

25. Johnston, I.G., Paladino, T., Gurd J.W. and Brown, I.R. (1990) Molecular cloning of SC1: A putative brain extracellular matrix glycoprotein showing partial similarity to osteonectin/BM40/SPARC. *Neuron* **2**, 165–176.
26. Kim, P.S. and Arvan, P. (1995) Calnexin and BiP act as sequential molecular chaperones during thyroglobulin folding in the endoplasmic reticulum. *J. Cell Biol.* **128**, 29–38.
27. Lane, T.F. and Sage, E.H. (1990) Functional mapping of SPARC: Peptides from two distinct Ca^{+2}- binding sites modulate cell shape. *J. Cell Biol.* **111**, 3065–3076.
28. Lane, T.F. and Sage, E.H. (1994) The biology of SPARC, a protein that modulates cell-matrix interactions. *FASEB (Fed. Amer. Soc. Exp. Biol) J.* **8**, 163–173.
29. Lankat-Buttgereit, B., Mann, K., Deutzmann, R., Timpl, R. and Krieg, T. (1988) Cloning and complete amino acid sequences of human and murine basement membrane protein BM-40 (SPARC/osteonectin). *FEBS (Fed. Eur. Biochem. Soc.) Lett.* **236**, 352–356.
30. Mann, K., Deutzman, R., Paulsson, M. and Timpl, R. (1987) Solubilization of protein BM-40 from a basement membrane tumor with chelating agents and evidence for its identity with osteonectin and SPARC. *FEBS (Fed. Eur. Biochem. Soc.) Lett.* **218**, 167–172.
31. Mason, I.J., Murphy, D., Munke, M., Francke, U., Elliott, R.W. and Hogan, B.L. (1986a) Developmental and transformation-sensitive expression of the Sparc gene on mouse chromosome 11. *EMBO (Eur. Mol. Biol. Organ.) J.* **5**, 1831–1837.
32. Mason, I.J., Taylor, A., Williams, J.G., Sage, H. and Hogan, B.L.M. (1986b) Evidence from molecular cloning that SPARC, a major product of mouse embryo parietal endoderm, is related to an endothelial cell 'culture shock' glycoprotein of Mr 43,000. *EMBO (Eur. Mol. Biol. Organ.) J.* **5**, 1465–1472.
33. Maurer, P., Mayer, U., Bruch, M., Jeno, P., Mann, K., Landwehr, R., Engel, J. and Timpl, R. (1992) High and low affinity calcium binding and stability of the multi-domain extracellular glycoprotein BM-40/SPARC/osteonectin. *Eur. J. Biochem.* **205**, 233–240.
34. Mayer, U., Aumailley, M., Mann, K., Timpl, R. and Engel, J. (1991) Calcium-dependent binding of basement membrane protein BM-40 (SPARC) to basement membrane collagen type IV. *Eur. J. Biochem.* **198**, 141–150.
35. McVey, J.H., Nomura, S., Kelly, P., Mason, I.J. and Hogan, B.L.M. (1988) Characterization of the mouse SPARC/osteonectin gene: Intron/exon organization and an unusual promoter region. *J. Biol. Chem.* **263**, 11111–11116.
36. Means, A.R., and Rasmussen, C.D. (1988) Calcium, calmodulin, and cell proliferation. *Cell Calcium* **9**, 313–319.
37. Mundlos, S., Schwahn, B., Reichert, T. and Zabel, B. (1992) Distribution of osteonectin mRNA and protein during human embryonic and fetal development. *J. Histochem. Cytochem.* **40**, 283–291.
38. Murano, S., Thweatt, R., Shmookler-Reis, R.J., Jones, R.A., Moerman, E.J. and Goldstein, S. (1992) Diverse gene sequences are overexpressed in Werner syndrome fibroblasts undergoing premature replicative senescence. *Mol. Cell. Biol.* **11**, 3905–3914.

39. Murphy-Ullrich, J.E., Lane, T.F., Pallero, M.A. and Sage, E.H. (1995) SPARC mediates focal adhesion disassembly in endothelial cells through a follistatin-like region and the Ca^{+2}-binding EF-hand. *J. Cell. Biochem.* **57**, 341–350.
40. Neri, M., Cancedda-Descalzi, F. and Cancedda, R. (1992) Heat shock response in cultured chick embryo chondryocytes. Osteonectin is a secreted heat-shock protein. *Eur. J. Biochem.* **205**, 569–574.
41. Ng, K.W., Manji, S.S., Young, M.F. and Findlay, D.M. (1989) Opposing influences of glucocorticoid and retinoic acid on transcriptional control in preosteoblasts. *Mol. Endocrinol.* **3**, 2079–2085.
42. Nischt, R., Pottgiesser, J., Krieg, T., Mayer, U., Aumailley, M. and Timpl, R. (1991) Recombinant expression and properties of the human calcium-binding extracellular matrix protein BM-40. *Eur. J. Biochem.* **200**, 529–536.
43. Nomura, S., Hasmi, S., McVey, J.H., Ham, J., Parker, M. and Hogan, B.L.M. (1989) Evidence for positive and negative regulatory elements in the 5′-flanking sequence of the mouse Sparc (osteonectin) gene. *J. Biol. Chem.* **264**, 12201–12207.
44. Otsuka, K., Yao, K.L., Wasi, S., Tung, P.S., Aubin, J.E., Sodek, J. and Termine, J.D. (1984) Biosynthesis of osteonectin by fetal porcine calvarial cells in vitro. *J. Biol. Chem.* **259**, 9805–9812.
45. Pober, J.S. (1988) Cytokine-mediated activation of vascular endothelium. Physiology and pathology. *Am. J. Pathol.* **133**, 426-433.
46. Porter, P.L., Sage, E.H., Lane, T.F., Funk, S.E. and Gown, A.M. (1995) Distribution of SPARC in normal and neoplastic human tissue. *J. Histochem. Cytochem.*, **43,** 791–800.
47. Pottgiesser, J., Maurer, P., Mayer, U., Nischt, R., Mann, K., Timpl, R., Krieg, T. and Engel, J. (1994) Changes in calcium and collagen IV binding caused by mutations in the EF hand and other domains of extracellular matrix protein BM-40 (SPARC,osteonectin). *J. Mol. Biol.* **238**, 563–574.
48. Reed, M.J., Puolakkainen, P., Lane, T.F., Dickerson, D., Bornstein, P. and Sage, E.H. (1993) Differential expression of SPARC and thrombospondin 1 in wound repair: immunolocalization and in situ hybridization. *J. Histochem. Cytochem.* **41**, 1467–1477.
49. Ringuette, M., Damjanovski, S. and Wheeler, D. (1991) Expression of SPARC/osteonectin in tissues of bony and cartilaginous vertebrates. *Biochem. Cell Biol.* **69**, 245–250.
50. Romberg, R.W., Werness, P.G., Lollar, P., Riggs, B.L. and Mann, K.G. (1985) Isolation and characterization of native adult osteonectin. *J. Biol. Chem.* **260**, 2728–2736.
51. Sage, E.H., Johnson, C. and Bornstein, P. (1984) Characterization of a novel serum albumin-binding glycoprotein secreted by endothelial cells in culture. *J. Biol. Chem.* **259**, 3993–4007.
52. Sage, E.H. (1992) Modulation of endothelial cell shape by SPARC does not involve chelation of extracellular Ca^{+2} and Mg^{+2}. *Biochem. Cell Biol.* **70**, 56–62.
53. Sage, E.H., Bassuk, J.A., Yost, J.C., Folkman, M.J. and Lane, T.F. (1995) Inhibition of endothelial cell proliferation by SPARC is mediated through a Ca^{+2}-binding EF-hand sequence. *J. Cell. Biochem.* **57**, 127–140.

54. Sage, E.H. and Bornstein, P. (1991) Extracellular proteins that modulate cell-matrix interactions: SPARC, tenascin, and thrombospondin. *J. Biol. Chem.* **266**, 14831–14834.
55. Sage, H., Tupper, J. and Bramson, R. (1986) Endothelial cell injury in vitro is associated with increased secretion of an Mr 43,000 glycoprotein ligand. *J. Cell Physiol.* **127**, 373–387.
56. Sage, H., Vernon, R.B., Funk, S.E., Everitt, E.A. and Angello, J. (1989) SPARC, a secreted protein associated with cellular proliferation, inhibits cell spreading in vitro and exhibits Ca^{+2}-dependent binding to the extracellular matrix. *J. Cell Biol.* **109**, 341–356.
57. Schnitzer, J.E. and Oh, P. (1992) Antibodies to SPARC inhibit albumin binding to SPARC, gp60, and microvascular endothelium. *Am. J. Physiol.* **263**, 1872–1879.
58. Schwarzbauer, J.E. and Spencer, C.S. (1993) The Caenorhabditis elegans homologue of the extracellular calcium binding protein SPARC/osteonectin affects nematode body morphology and mobility. *Mol. Biol. Cell* **4**, 941–952.
59. Swaroop, A., Hogan, B.L.M. and Franke, U. (1988) Molecular analysis of the cDNA for human SPARC/osteonectin/BM-40: sequence, expression, and localization of the gene to chromosome 5q31-q33. *Genomics* **2**, 37–47.
60. Termine, J.D., Kleinman, H.K., Whitson, S.W., Conn, K.M., McGarvey, M.L. and Martin, G.R. (1981) Osteonectin a bone-specific protein linking mineral to collagen. *Cell* **26**, 99–105.
61. Tremble, P.M., Lane, T.F., Sage, E.H. and Werb, Z. (1993) SPARC, a secreted protein associated with morphogenesis and tissue remodeling, induces expression of metalloproteinases in fibroblasts through a novel extracellular matrix-dependent pathway. *J. Cell Biol.* **121**, 1433–1444.
62. Villarreal, X.C., Mann, K.G. and Long, G.L. (1989) Structure of human osteonectin based upon analysis of cDNA and genomic sequence. *Biochemistry* **28**, 6483–6491.
63. Wewer, U.M., Albrechtsen, R., Fisher, L.W., Young, M.F. and Termine, J.D. (1988) Osteonectin/SPARC/BM-40 in human decidua and carcinoma, tissues characterized by de novo formation of basement membrane. *Am. J. Pathol.* **132**, 345–355.
64. Yost, J.C., Bell, A., Seale, R. and Sage, E.H. (1994) Purification of biologically active SPARC expressed in Saccharomyces cerevisiae. *Arch. Biochem. Biophys.* **314**, 50–63.
65. Yost, J.C. and Sage, E.H. (1993) Specific interaction of SPARC with endothelial cells is mediated through a carboxyl-terminal sequence containing a calcium-binding EF hand. *J. Biol. Chem.* **268**, 25790–25796.
66. Young, M.F., Findlay, D.M., Dominguez, P., Burbelo, P.D., McQuillan, C., Kopp, J.B., Gehron Robey, P., Hefferan, T.E. and Termine, J.D. (1989) Osteonectin promoter: DNA sequence analysis and S1 endonuclease site potentially associated with transcriptional control in bone cells. *J. Biol. Chem.* **264**, 450–456.

7. COUNTERADHESIVE PROPERTIES OF CHONDROITIN SULFATE PROTEOGLYCANS

RICHARD U. MARGOLIS and RENÉE K. MARGOLIS

Department of Pharmacology, New York University Medical Center, New York, NY 10016, and Department of Pharmacology, State University of New York, Health Science Center, Brooklyn, NY 11203

It is obvious that as a corollary to cell-cell and cell-matrix adhesion, mechanisms must also exist which mediate detachment, and thereby permit cell migration which is essential for normal developmental processes as well as for tumor metastasis. Chondroitin sulfate proteoglycans are one class of macromolecules that has recently been recognized to have counteradhesive effects which in many cases are probably of physiological significance. While there has been considerable interest in the ability of heparan sulfate proteoglycans to mediate cell adhesion via so-called heparin binding sequences present in a variety of cell surface and extracellular matrix proteins, less work has been devoted to the effects on cell adhesion of chondroitin sulfate proteoglycans. Tenascin, which exists to a small extent in the form of a "part-time" chondroitin sulfate proteoglycan (Hoffman *et al.*, 1988; Grumet *et al.*, 1994), is the subject of several other chapters in this volume and will therefore not be covered here. Some early studies also showed that high concentrations of cartilage aggrecan inhibited the attachment of fibroblasts to tissue culture plastic, collagen, and fibronectin (Knox and Wells, 1979; Rich *et al.*, 1981), but these results may be attributable to relatively nonspecific effects of the high negative charge density of the substrate. With respect to nervous tissue, it was reported that added glycosaminoglycans or glycosaminoglycan-coated substrata modified the adhesion and neurite extension of cultured PC12 cells (Akeson and Warren, 1986; Damon *et al.*, 1988) and dorsal root ganglia (Carbonetto *et al.*, 1983; Verna *et al.*, 1989). In our own studies we have also found that β-xylosides have significant effects on the adhesion and morphology of PC12 pheochromocytoma cells and primary cultures of rat cerebellum (Margolis *et al.*, 1991). Chondroitin sulfate and other sulfated glycosaminoglycans are *O*-glycosidically linked to serine residues in a core protein through a characteristic -glucuronosyl-galactosyl-galactosyl-xylosyl-*O*-serine linkage region. β-Xylosides competitively inhibit the synthesis of glycosaminoglycan chains on xylosylated core protein, resulting in carbohydrate-deficient proteoglycans which are usually accompanied by large amounts of free glycosaminoglycans in which chain initiation occurs on the exogenous β-xyloside acceptor. β-Xyloside treatment resulted in up to 11-fold increases in [^{35}S]sulfate-labeled macromolecules (mostly chondroitin sulfate) released into the culture medium, accompanied by an increase in length and decrease in the branching, diameter, and adhesion to the collagen substratum of PC12 cell processes. There were also distinct alterations in the adhesion and morphology of monolayer, aggregate, and explant cultures of early postnatal rat cerebellum, which occurred together with inhibition of chondroitin sulfate proteoglycan biosynthesis and a decrease in glycosaminoglycan chain size (Margolis *et al.*, 1991). In this chapter we will briefly review recent work pertaining to nervous tissue as well as studies on the effects of chondroitin sulfate proteoglycans of known primary structure from other sources, all of which indicate that in distinction to the heparan sulfate proteoglycans, chondroitin sulfate proteoglycans generally have counteradhesive effects.

NEUROCAN AND PHOSPHACAN

Neurocan and phosphacan are two chondroitin sulfate proteoglycans which appear to be specific to nervous tissue, where they account for over half of the total chondroitin sulfate proteoglycan content of brain. Although they are structurally unrelated, we will discuss them together because they were both biochemically characterized and cloned in our laboratories, and they share several biological properties which we have described based on parallel assays using identical experimental approaches.

Neurocan, which is synthesized by neurons, has a 138 kDa core protein and three chondroitin sulfate chains which during the course of postnatal brain development increase in size (from 22 to 32 kDa) and 4- versus 6-sulfation (80%/20% to 97%/<3%) (Rauch *et al.*, 1991, 1992). Near its N-terminus neurocan contains tandem repeats characteristic of other hyaluronic acid-binding chondroitin sulfate proteoglycans such as aggrecan and versican, and forms link protein-stabilized aggregates with hyaluronic acid. Its C-terminal portion contains EGF-like, lectin-like, and complement regulatory protein-like domains similar to those found (but in a different order) in the selectins, and these N-terminal and C-terminal homology domains are separated by a central domain, accounting for ~50% of the amino acids, having no homology to other reported protein sequences (Rauch *et al.*, 1992). The concentration of neurocan in brain increases during late embryonic development but then declines steeply together with hyaluronic acid during the early postnatal period, and neurocan also undergoes extensive proteolytic processing during the course of brain development (Rauch *et al.*, 1991; Margolis and Margolis, 1994). It is possible that one function of neurocan and similar hyaluronic acid-binding chondroitin sulfate proteoglycans such as versican and brevican (Figure 7.1) may be to reversibly link the cell surface and the extracellular matrix in the developing central nervous system.

Phosphacan is synthesized by astroglia and contains a 173 kDa core protein consisting of an N-terminal carbonic anhydrase-like domain followed by a single fibronectin type III sequence, and three to four 28 kDa chondroitin

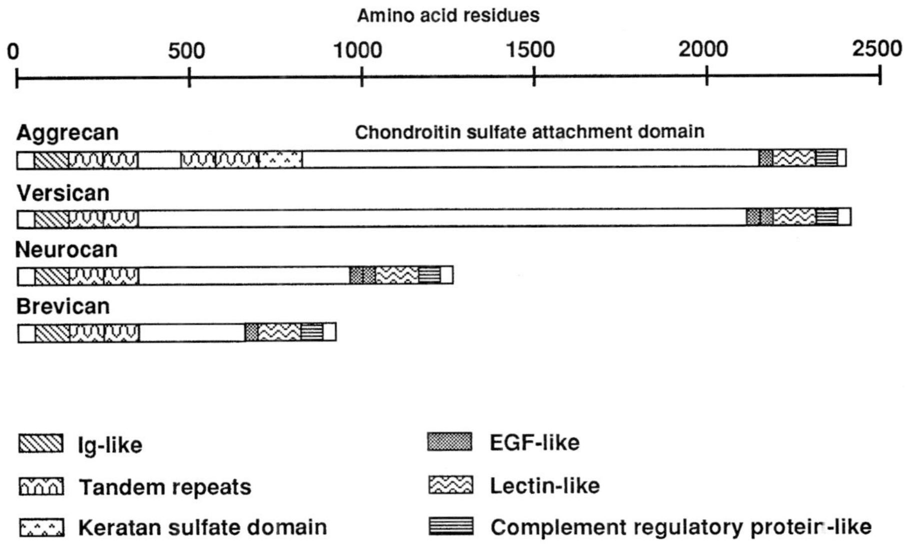

Figure 7.1 Primary structures and homology domains of aggrecan, versican, neurocan and brevican.

sulfate chains which also show a developmental increase in 4- versus 6-sulfation similar to that of neurocan. Keratan sulfate-containing glycoforms of phosphacan (designated phosphacan-KS) are also present in postnatal brain (Rauch *et al.*, 1991). Cloning of phosphacan (Maurel *et al.*, 1994) revealed that it is an alternative splicing product that contains the entire extracellular domain of a receptor-type protein tyrosine phosphatase (RPTP-ζ/β), which also occurs as a chondroitin sulfate proteoglycan in brain (Shitara *et al.*, 1994).

In initial experiments using a gravity cell binding assay, we found that low concentrations of neurocan or phosphacan inhibited the binding of neurons to the neuron-glial cell adhesion molecule Ng-CAM/L1 when mixtures containing the proteoglycan and Ng-CAM were adsorbed to polystyrene dishes, whereas cartilage aggrecan had relatively little effect (Grumet *et al.*, 1993). Both native neurocan and phosphacan or their core glycoproteins obtained by chondroitinase treatment were effective in this assay, and cell binding was inhibited in a concentration-dependent manner, with essentially complete inhibition being obtained at a proteoglycan concentration of 10 μg/ml when mixed with the same concentration of Ng-CAM. Low concentrations of these brain proteoglycans also inhibited the homophilic self-aggregation of microbeads coated with the neural cell adhesion molecules Ng-CAM and N-CAM, but it would appear that the proteoglycans selectively affect only certain adhesive systems, insofar as there was very little inhibition of neuronal binding to laminin.

Using a radioligand binding assay, we later directly demonstrated that neurocan and phosphacan bound to Ng-CAM/L1/NILE and N-CAM, but not to over a dozen other cell surface and extracellular matrix proteins tested. The binding was saturable, reversible, and of high affinity, with a K_d of ~0.1 nM (Friedlander *et al.*, 1994; Milev *et al.*, 1994). Phosphacan binding was reduced by ~15% following chondroitinase treatment and free chondroitin sulfate was only moderately inhibitory, indicating that the core glycoprotein accounts for most of its binding activity, whereas a large portion of the neurocan binding requires the presence of chondroitin sulfate chains. Neurocan and phosphacan were also found to be potent inhibitors of neurite outgrowth (Friedlander *et al.*, 1994; Milev *et al.*, 1994).

In view of the binding of neurocan and phosphacan to Ng-CAM, we also examined the effects of these proteoglycans on neuron binding to *a substrate* consisting of anti-Ng-CAM antibodies (to which the proteoglycans do not bind), rather than to Ng-CAM itself (Friedlander *et al.*, 1994; Milev *et al.*, 1994). Similar results were obtained in these experiments, indicating that neurocan and phosphacan have a *direct* inhibitory effect on cell binding independent of their ability to bind to substrate Ng-CAM.

Although in longer term (>1 h) gravity adhesion and neurite growth assays, neurocan and phosphacan inhibited cell binding, in a short term (~1 min) centrifugation assay neurons bound to neurocan, phosphacan, and phosphacan-KS (Grumet *et al.*, 1993; Milev *et al.*, 1994). Attachment of cells in the longer term assays involves interactions of the coated proteins with the cell surface to induce cellular responses such as cell spreading. In contrast, the short term centrifugation assay does not require cellular responses for adhesion, and it probably provides a measure only of the initial interaction between the cell membrane and the coated protein. Thus, the lack of cell adhesion and neurite growth in the presence of neurocan and phosphacan in the gravity assay is likely to be due to inability of the cells to establish stable contacts after initial binding to the substrate.

Neurocan and phosphacan also bound saturably and reversibly to tenascin, but with an affinity (K_d= ~3 nM) that was somewhat lower than that determined for their binding to the neural cell adhesion molecules (Grumet *et al.*, 1994). Substrate-bound tenascin is known to have effects on different cell types (Crossin, Prieto and Mauro, this volume), including promoting adhesion of human U-251 glioma cells (Bourdon and Ruoslahti, 1989). Because we found that rat C6 glioma cells bound to tenascin and extended processes, we investigated whether interactions between tenascin and brain proteoglycans affected the adhesion of these cells (Grumet *et al.*, 1994). C6 glioma cells bound to polystyrene dishes coated with tenascin or laminin but not to dishes coated with phosphacan or neurocan. When tenascin or laminin coated areas were treated with a second coating of neurocan there was little or no effect on cell binding, whereas a second coating with phosphacan produced a concentration dependent inhibition of binding to tenascin but not to laminin. Using ^{125}I-labeled proteoglycans it was shown that neurocan and phosphacan bound equally well to tenascin coated polystyrene dishes. When the order of protein coating was reversed and the proteoglycans were used as a first coat, a second treatment with tenascin allowed adhesion of C6 glioma cells, but higher concentrations of tenascin on phosphacan were required to obtain levels of adhesion equal to those obtained on neurocan. These results indicate that phosphacan binds to tenascin in a manner that can inhibit interactions of C6 glioma cells with tenascin.

THE NG2 PROTEOGLYCAN

The antigen recognized by the NG2 monoclonal antibody is an ~500 kDa integral membrane chondroitin sulfate proteoglycan with a 252 kDa core protein consisting of a large extracellular domain (2,224 amino acids), a single transmembrane domain, and a short cytoplasmic tail (Nishiyama *et al.*, 1991a). The extracellular domain contains four internal repeats, each consisting of 200 amino acids and containing a short sequence that resembles the putative calcium-binding region of the cadherins. However, the major portion of the sequence has no significant similarity to other known proteins.

The NG2 proteoglycan was originally identified on rat neural cell lines with characteristics of both neuronal and glial cells, and is found in rat optic nerve and cerebellum on O2A glial progenitor cells which have the potential to differentiate *in vitro* into oligodendrocytes or type 2 astrocytes (Stallcup and Beasley, 1987; Levine and Stallcup, 1987). *In vivo* NG2 has been found on stellate, process-bearing cells in the rat cerebellum, but not on cells with morphological characteristics of mature oligodendrocytes (Levine and Card, 1987), suggesting that its expression is turned off before the terminal differentiation of the progenitors into oligodendrocytes. Immunocytochemical studies of developing rat embryos have also shown a widespread distribution in non-neural tissues, especially in the developing mesenchyme, and the expression of NG2 appears to be developmentally regulated insofar as the level of expression is highest on immature proliferating cells and decreases as these cells differentiate (Nishiyama *et al.*, 1991b).

Antibodies to NG2 coimmunoprecipitate type VI collagen from some cell lines, and the binding between the two molecules appears to be mediated by protein-protein interactions (Stallcup *et al.*, 1990). Treatment of these cells resulting in a change in the distribution of NG2 on the cell surface also causes a parallel change in the distribution of type VI collagen, suggesting that NG2 may function as a cell surface receptor for collagen VI and mediate cell-matrix interactions by this mechanism (Stallcup *et al.*, 1990; Nishiyama and Stallcup, 1993). The unusual class of glia in rat cerebellum that express the NG2 proteoglycan also appear to be a reactive cell type that responds to brain injury, insofar as it has been shown that there is an increased expression of the proteoglycan in the region of small puncture lesions of rat cerebellum (Levine, 1994).

It was recently reported that the NG2 proteoglycan, when adsorbed onto nitrocellulose-coated surfaces, does not support attachment and neurite outgrowth of cerebellar granule neurons (which do not attach to nitrocellulose alone but do bind to laminin and L1), and inhibits neurite outgrowth on laminin-coated surfaces (Dou and Levine, 1994). The question of whether the NG2 proteoglycan was capable of inhibiting neuronal adhesion when applied in the form of *mixtures* with laminin or L1 was not addressed in these studies, but it is interesting that as in the case of phosphacan, the NG2 core protein obtained by chondroitinase treatment retained nearly full effectiveness (as compared to the intact proteoglycan) in inhibiting neurite outgrowth. Elongating axons also avoided the NG2 substrate at borders of NG2 and laminin or L1.

VERSICAN

Versican is a large, multidomain, hyaluronic acid-binding chondroitin sulfate proteoglycan which is structurally related to aggrecan, neurocan, and brevican (Fig. 7.1). It was originally cloned from a human fibroblast cDNA library, and shown to contain a 2389 amino acid core protein with an N-terminal domain homologous to the hyaluronic acid-binding domains of link protein and cartilage aggrecan, a C-terminal domain containing two EGF-like repeats, a lectin-like sequence, and a complement regulatory protein-like domain, and a central nonhomologous region containing up to 15 potential attachment sites for chondroitin sulfate chains (Zimmermann and Ruoslahti, 1989). It has also recently been shown that there is alternative splicing of human versican insofar as its core protein may contain either or both or two central chondroitin sulfate attachment domains of unequal size (Dours-Zimmermann and Zimmermann, 1994). The largest form (V0) is present in a variety of tissues including nervous tissue, whereas keratinocytes expressed exclusively the intermediate size isoform (V1), and the short versican variant (V2) was detected only in brain. It is also noteworthy that the core proteins of brevican, neurocan, and the three splice variants of versican form an even array with chondroitin sulfate attachment domains of approximately 300, 600, 1000, 1750, and 2750 amino acids, suggesting that variations in the chondroitin sulfate content of these structurally similar proteoglycans may have functional significance.

A related chondroitin sulfate proteoglycan from chicken (originally designated PG-M) has also been cloned (Shinomura *et al.*, 1993). It has a 388 kDa core protein which is expressed during chondrogenesis in chick limb buds, as well as in a number of other chick embryo cells and tissues including fibroblasts, aorta, skin, lung, and muscle, and appears to be the chick homolog of versican. In chick this proteoglycan occurs in two alternatively spliced forms, the larger of which is ~100 kDa greater in size than human versican. It has 68–82% amino acid sequence identity with human versican in its three

N-terminal domains, 68–84% identity in the two EGF-like domains, and 93–96% identity in the lectin-like and complement regulatory protein-like sequences, but only 38–44% identity in the central chondroitin sulfate attachment region (where the alternative splicing occurs and which accounts for ~80% of its size). Multiple forms of mouse versican (PG-M) with a similarly high degree of sequence identity to human versican have also recently been described (Ito *et al.*, 1995). As in the case of human versican, there is alternative splicing of either or both of two chondroitin sulfate attachment domains which varies between tissues to yield four different core proteins, including one in which both domains have been deleted (Zako *et al.*, 1995).

When chick embryo versican (PG-M) was added to the culture medium, various types of trypsinized cells (including baby hamster kidney cells, chick embryo fibroblasts, Chinese hamster ovary cells, and melanoma cells) failed to adhere to substrates coated with fibronectin, collagen, or vitronectin, as well as to other substrates such as laminin and RGDS-derivatized BSA to which versican itself does not bind (Yamagata *et al.*, 1989). Treatment of the proteoglycan with either chondroitinase or proteases abolished its inhibitory properties, demonstrating that the chondroitin sulfate chains are necessary but not sufficient for this biological activity. If the adsorption of soluble versican to available plastic surfaces of previously coated dishes was blocked by treatment with BSA its inhibitory effects were abolished, suggesting that it is only the immobilized fraction of the proteoglycan that inhibits cell adhesion, and other studies indicated that the active site for the interaction between cells and versican resides on the chondroitin sulfate chains. It has been proposed that possible tissue-specific differences in the counteradhesive activity of versican may be a function of different amounts of chondroitin sulfate on the core protein, as a result of the tissue-dependent alternative splicing (Ito *et al.*, 1995).

The finding that a chondroitin sulfate proteoglycan synthesized by fibroblasts can inhibit adhesion of these cells to substrates such as fibronectin is consistent with models of cell migration in which it is postulated that heparan sulfate proteoglycans and fibronectin play a direct role in cell-substrate adhesion whereas chondroitin sulfate proteoglycans and hyaluronic acid are inhibitory (Culp *et al.*, 1978). The pertubation of neural crest cell (Perris and Johansson, 1987, 1990) and cardiac mesenchymal cell (Funderburg and Markwald, 1986) migration by added or coated chondroitin sulfate proteoglycans supports the concept that they may have a regulatory role in cell adhesion and migration.

DECORIN AND BIGLYCAN

Decorin and biglycan are small chondroitin sulfate or chondroitin/dermatan sulfate proteoglycans containing one or two ~40 kDa glycosaminoglycan chains at serine-5 or serine-5 and -11, respectively (for a review, see Kresse *et al.*, 1993). A survey of developing human tissues has shown that decorin is associated with fibrillar elements of all fetal connective tissues examined, whereas biglycan was found to be cell surface-associated or in the pericellular space of a wide variety of cell types including epidermal cells, blood vessel endothelial cells, kidney collecting tubule epithelial cells, skeletal myofibrils, bone cells, and chondrocytes, but not associated with classical collagen fibrils (Bianco *et al.*, 1990). Both decorin and biglycan have 37–38 kDa core proteins which are characterized by a series of 7–11 non-overlapping leucine-rich repeats that are thought to serve as collagen and fibronectin binding sites. The deduced protein sequences of rat (Dreher *et al.*, 1990), human (Fisher *et al.*, 1989), and bovine (Torok *et al.*, 1993) biglycan are 95–97% identical in the secreted protein, whereas there is only 77% identity between rat, human, bovine, and chicken decorin (Asundi and Dreher, 1992). Human, bovine, or rat decorin core proteins share 54–55% identity with biglycan from the same species (Fisher *et al.*, 1989; Neame *et al.*, 1989; Asundi and Dreher, 1992).

It was originally reported that a mixture of chondroitin/dermatan sulfate proteoglycans (decorin and biglycan) isolated from bovine articular cartilage inhibited the attachment and spreading of 3T3 cells on tissue culture plastic coated with plasma fibronectin, whereas cartilage chondroitin sulfate proteoglycan (aggrecan) had little or no effect (Lewandowska *et al.*, 1987). Dermal fibroblasts were much less sensitive to either class of proteoglycan, and substrates coated with either type of proteoglycan did not support cell attachment. Although the native intact proteoglycans produced moderate to complete inhibition of cell attachment when used at concentrations of 100–300 μg/ml, neither the free glycosaminoglycan chains nor the core proteins were effective. The proteoglycans did not inhibit attachment or spreading of 3T3 cells or dermal fibroblasts on substrata composed of either type I collagen alone or type I collagen to which fibronectin was adsorbed, indicating the existence of collagen-dependent, fibronectin-independent mechanisms of fibroblast adhesion, and that the collagen binding sites for these proteoglycans are different from those for cell surface "receptors". Affinity chromatography studies supported other evidence indicating that these effects involved binding of the proteoglycans to glycosaminoglycan-binding sites of substratum-bound fibronectin, including a cryptic site in the cell-binding domain, thereby inhibiting the interaction of fibronectin with cell surface receptors.

It was later shown that decorin purified under non-denaturing conditions from the medium of cultured human skin fibroblasts inhibited the adhesion of fibroblasts to a fibronectin substrate (Winnemöller *et al.*, 1991). Because the core protein itself was as effective as the intact proteoglycan, it was concluded that the inhibition was caused by an interaction of the core protein with the cell-binding domain of fibronectin. These authors also did not find differential effects of decorin on adhesion of dermal fibroblasts as compared to 3T3 cells. The decorin core protein domain that interacts with the cell-binding domain of fibronectin is located in the C-terminal disulfide loop of decorin, and does not contain the amino acid sequence interacting with the heparin-binding domains of fibronectin (Kresse *et al.*, 1993). It was suggested by Winnemöller *et al.* (1991) that the discrepancy between their results and those of Lewandowska *et al.* (1987) is probably due to a decrease in the functional activity of the core protein in decorin purified from cartilage using strong denaturing conditions.

In subsequent studies, both decorin and biglycan were reported to inhibit the initial attachment of CHO cells and rat embryo fibroblasts to substrates composed of the 105 kDa cell-binding fibronectin fragment, but did not affect cell attachment to intact fibronectin (Bidanset *et al.*, 1992). Although this effect involves proteoglycan binding to the substrate via the protein core, the intact proteoglycan is required and isolated core proteins are inactive. Moreover, the efficacies of the different proteoglycans in these assays do not appear to be related to the proportions of chondroitin sulfate and dermatan sulfate glycosaminoglycan chains (which are tissue-specific properties of decorin and biglycan; Choi *et al.*, 1989), since there was no relationship between the ability of the proteoglycans to inhibit cell attachment and their glucuronic/iduronic acid content. Neither the percentage of cells spread nor the mean area of spread cells adhering to intact fibronectin were affected by decorin or biglycan. However, significantly fewer cells formed focal adhesions in the presence of the proteoglycans as compared with untreated control cells, suggesting that the proteoglycan-fibronectin binding may affect several stages in the cell adhesion process. The reasons for the discrepancy between these results and those of earlier studies showing that decorin and biglycan inhibit cell attachment to intact fibronectin (Lewandowska *et al.*, 1987; Winnemöller *et al.*, 1991) as compared to the 105 kDa cell-binding fragment are not entirely clear, but it has been suggested that they may be related to differences in the techniques and cell types used (Bidanset *et al.*, 1992).

In contrast to these studies on the influence of decorin and biglycan on fibroblast attachment, overexpression of decorin in Chinese hamster ovary cells leads to an increased cell spreading and inhibits cell proliferation (Yamaguchi and Ruoslahti, 1988). However, these effects may be mediated at least in part by transforming growth factor-β which both binds to decorin and stimulates its biosynthesis (Yamaguchi *et al.*, 1990), rather than representing a direct result of decorin interaction with the substrate.

Decorin also binds with high affinity to thrombospondin and inhibits attachment of fibroblasts to a thrombospondin substrate (Winnemöller *et al.*, 1992). Because of the presence of multiple binding sites on both decorin and thrombospondin the respective domains involved in these interactions have not been identified.

The biological significance of the decorin-mediated inhibition of fibroblast adhesion is still unclear, but decorin is known to be a component of a feedback system regulating cell growth and the composition of the extracellular matrix, and some in vitro studies suggest that these inhibitory effects may play a role in processes such as wound healing.

SERGLYCIN

Serglycin is a chondroitin sulfate proteoglycan that was originally identified in a yolk sac tumor, and was later found in a variety of blood cells including mast cells, basophilic leukemia cells, platelets, NK-cells, macrophage-like cells, and eosinophils (for references, see Kjellén and Lindahl, 1990). It has a 17–19 kDa core protein with up to 24 consecutive serine-glycine repeats which are heavily substituted with glycosaminoglycans. These include not only chondroitin sulfate but also dermatan sulfate, heparan sulfate, and heparin, depending on the cell type.

Serglycin was found to inhibit the adhesion and spreading of rat yolk sac tumor cells to fibronectin and collagen type I (Brennan *et al.*, 1983). The proteoglycan was shown to bind 12-times as well to type I as compared to type IV collagen, to which it did not inhibit cell attachment, nor did it inhibit cell attachment to fibronectin fragments which lack the heparin-binding site. Based on these data, it was concluded that serglycin did not inhibit cell adhesion by competition with cell surface proteoglycan binding to collagen or fibronectin, but rather by hindering the interaction of the cell surface with cell attachment sites on the substrate (either sterically or by electrostatic repulsion) as a function of the physical properties of the large, bulky serglycin molecule bound to fibronectin via its glycosaminoglycan chains. This effect of serglycin on yolk sac tumor cell adhesion therefore differs from the effects of proteoglycans on CHO cell attachment to either intact fibronectin or the 105 kDa cell-binding fragment, insofar as they were not affected by cartilage aggrecan at a concentration of 100 μg/ml whereas decorin and biglycan showed significant inhibition as concentrations as low as 10 μg/ml (Bidanset *et al.*, 1992).

It has been suggested that chondroitin sulfate proteoglycans produce a focal reduction of cell adhesion (Culp *et al.*, 1979), and it was hypothesized by Brennan and coworkers that such inhibitory effects may be particularly prominent in tumors which have been reported to produce increased amounts of chondroitin sulfate proteoglycans. In this manner, serglycin and other chondroitin sulfate proteoglycans might modulate the invasive movements of tumor cells by decreasing cell interactions with extracellular matrix components such as fibronectin and type I collagen while not affecting their attachment to type IV collagen in basement membranes (Brennan *et al.*, 1983).

CONCLUSIONS

The literature reviewed above shows a considerable range of effects of chondroitin sulfate proteoglycans on cell adhesion, which is not surprising insofar as these studies cover a wide variety of structurally diverse proteoglycans, and of cell types and experimental conditions. Although depending on the type of assay or substrate used, even studies of the same proteoglycan and cells may yield partially conflicting results, the most notable conclusion which one may draw from the still very limited number of studies in this area is that chondroitin sulfate proteoglycans generally have counteradhesive effects on cells.

These proteoglycans are, for the most part, present in the extracellular matrix. However NG2, whose concentration in brain appears to be considerably lower than that of other chondroitin sulfate proteoglycans such as neurocan and phosphacan, is an integral membrane protein, and phosphacan is present in nervous tissue both as an extracellular proteoglycan as a result of alternative splicing, together with much lower levels of the full-length transmembrane protein tyrosine phosphatase, which also occurs in the form of a chondroitin sulfate proteoglycan. Soluble isoforms of various membrane proteins have been predicted to be generated by alternative splicing (for references, see Ehlers and Riordan, 1991; Fernandez-Botran, 1991). Although in many cases there is no direct evidence at the molecular level that the spliced messages can actually be translated and properly processed within cells, it was recently shown that an alternatively spliced isoform of P-selectin is present *in vivo* as a soluble molecule (Ishiwata *et al.*, 1994). The biological significance of generating a soluble extracellular form of a receptor-type phosphatase is not yet clear, but our studies suggest that by binding to neural cell adhesion molecules and extracellular matrix proteins such as tenascin, and possibly also by competing for ligands of the transmembrane phosphatase, phosphacan may modulate cell-cell and cell-matrix interactions in the developing central nervous system.

The generally counteradhesive effects of the chondroitin sulfate proteoglycans stand in contrast to the heparan sulfate proteoglycans, which are usually integral membrane proteins present on the cell surface, or a component of basement membranes. In view of the heparan sulfate/heparin binding sites present on a number of cell surface and extracellular matrix proteins such as N-CAM, laminin, and fibronectin, heparan sulfate proteoglycans are generally considered to be involved in the initiation and maintenance of cell-matrix and cell-cell adhesion (Reyes *et al.*, 1990; Esko, 1991; Bernfield *et al.*, 1992; David, 1993; Stanley *et al.*, 1995), in contrast to the counteradhesive effects of the chondroitin sulfate proteoglycans described above. Although the details of these effects of chondroitin sulfate proteoglycans still remain to be elucidated, including such aspects as the relative roles of the core protein as compared to the glycosaminoglycan chains and the influence of proteoglycan size and structure, at the risk of very considerable oversimplification one can interpret the available data as indicating that the chondroitin sulfate and heparan sulfate proteoglycans may have opposing effects on adhesive processes. Further work will undoubtedly fill in the many missing pieces in our current picture of these interactions, and may also provide new information that argues against the generalization tentatively advanced above.

REFERENCES

1. Akeson, R. and Warren, S.L. (1986) PC12 adhesion and neurite formation on selected substrates are inhibited by some glycosaminoglycans and a fibronectin-derived tetrapeptide. *Exp. Cell Res.* **162**: 347–362.
2. Asundi, V.K. and Dreher, K.L. (1992) Molecular characterization of vascular smooth muscle decorin: Deduced core protein structure and regulation of gene expression. *Eur. J. Cell Biol.* **59** 314–321.
3. Bernfield, M., Kokenyesi, R., Kato, M., Hinkes, M.T., Spring, J., Gallo, R.L., and Lose, E.J. (1992) Biology of the syndecans: A family of transmembrane heparan sulfate proteoglycans. *Annu. Rev. Cell Biol.* **8**: 365–393.
4. Bianco, P., Fisher, L.W., Young, M.F., Termine, J.D., and Robey, P.G. (1990) Expression and localization of the two small proteoglycans biglycan and decorin in developing human skeletal and non-skeletal tissues. *J. Histochem. Cytochem.* **38**: 1549–1563.
5. Bidanset, D.J., LeBaron, R., Rosenberg, L., Murphy-Ullrich, J.E. and Hook, M. (1992) Regulation of cell substrate adhesion: Effects of small galactosaminoglycan-containing proteoglycans. *J. Cell Biol.* **118**: 1523–1531.
6. Bourdon, M.A. and Ruoslahti, E. (1989) Tenascin mediates cell attachment through an RGD-dependent receptor. *J. Cell Biol.* **108**: 1149–1155.
7. Brennan, M.J., Oldberg, Å., Hayman, E.G., and Ruoslahti, E. (1983) Effect of a proteoglycan produced by rat tumor cells on their adhesion to fibronectin-collagen substrata. *Cancer Res.* **43**: 4302–4307.
8. Carbonetto, S., Gruver, M.M. and Turner, D.C. (1983) Nerve fiber growth in culture on fibronectin, collagen, and glycosaminoglycan substrates. *J. Neurosci.* **3**: 2324–2335.
9. Choi, H.U., Johnson, T.L., Pal, S., Tang, L.-H., and Rosenberg, L. (1989) Characterization of the dermatan sulfate proteoglycans, DS-PGI and DS-PGII, from bovine articular cartilage and skin isolated by octyl-sepharose chromatography. *J. Biol. Chem.* **264**: 2876–2884.
10. Culp, L.A., Rollins, B.J., Buniel, J., and Hitri, S. (1978) Two functionally distinct pools of glycosaminoglycan in the substrate adhesion site of murine cells. *J. Cell Biol.* **79**: 788–801.
11. Culp, L.A., Murray, B.A., and Rollins, B.J. (1979) Fibronectin and proteoglycans as determinants of cell-substratum adhesion. *J. Supramol. Struct.* **11**: 401–427.
12. Damon, D.H., D'Amore, P.A. and Wagner, J.A. (1988) Sulfated gylcosaminoglycans modify growth factor-induced neurite outgrowth in PC12 cells. *J. Cellular Physiol.* **135**: 293–300.
13. David, G. (1993) Integral membrane heparan sulfate proteoglycans. *FASEB J.* **7**: 1023–1030.
14. Dou, C.-L. and Levine, J.M. (1994) Inhibition of neurite growth by the NG2 chondroitin sulfate proteoglycan. *J. Neurosci.* **14**: 7616–7628.
15. Dours-Zimmermann, M.T. and Zimmermann, D.R. (1994) A novel glycosaminoglycan attachment domain identified in two alternative splice variants of human versican. *J. Biol. Chem.* **269**: 32992–32998.
16. Dreher, K.L., Asundi, V., Matzura, D., and Cowan, K. (1990) Vascular smooth muscle biglycan represents a highly conserved proteoglycan within the arterial wall. *Eur. J. Cell Biol.* **53**: 296–304.

17. Ehlers M.R.W. and Riordan J.F. (1991). Membrane proteins with soluble counterparts: Role of proteolysis in the release of transmembrane proteins. *Biochemistry* **30**: 10065–10074.
18. Esko, J. D. (1991) Genetic analysis of proteoglycan structure, function and metabolism. *Curr. Opin. Cell Biol.* **3**: 805–816.
19. Fernandez-Botran R. (1991). Soluble cytokine receptors: their role in immunoregulation. *FASEB J.* **5**: 2567–2574.
20. Fisher, L.W., Termine, J.D., and Young, M.F. (1989) Deduced protein sequence of bone small proteoglycan I (Biglycan) shows homology with proteoglycan II (Decorin) and several nonconnective tissue proteins in a variety of species. *J. Biol. Chem.* **264**: 4571–4576.
21. Friedlander, D., Milev, P., Karthikeyan, L., Margolis, R.K., Margolis, R.U., and Grumet, M. (1994) The neuronal chondroitin sulfate proteoglycan neurocan binds to the neural cell adhesion molecules Ng-CAM/L1/NILE and N-CAM and inhibits neuronal adhesion and neurite outgrowth. *J. Cell Biol.* **125**: 669–680.
22. Funderburg, F.M. and Markwald, R.R. (1986) Conditioning of native substrates by chondroitin sulfate proteoglycans during cardiac mesenchymal cell migration. *J. Cell Biol.* **103**: 2475–2487.
23. Grumet, M., A. Flaccus, and R.U. Margolis. (1993) Functional characterization of chondroitin sulfate proteoglycans of brain: Interactions with neurons and neural cell adhesion molecules. *J. Cell Biol.* **120**: 815–824.
24. Grumet, M., Milev, P., Sakurai, T., Karthikeyan, L., Bourdon, M., Margolis, R.K., and Margolis, R.U. (1994) Interactions with tenascin and differential effects on cell adhesion of neurocan and phosphacan, two major chondroitin sulfate proteoglycans of nervous tissue. *J. Biol. Chem.* **269**: 12142–12146.
25. Hoffman, S., Crossin, K.L., and Edelman, G.M. (1988) Molecular forms, binding functions, and developmental expression patterns of cytotactin and cytotactin-binding proteoglycan, an interactive pair of extracellular matrix molecules. *J. Cell Biol.* **106**: 519–532.
26. Ishiwata N., Takio K., Katayama M., Watanabe K., Titani K., Ikeda Y. and Handa M. (1994). Alternatively spliced isoform of P-selectin is present *in vivo* as a soluble molecule. *J. Biol. Chem.* **269**: 23708–23715.
27. Ito, K., Shinomura, T., Zako, M., Ujita, M., and Kimata, K. (1995) Multiple forms of mouse PG-M, a large chondroitin sulfate proteoglycan generated by alternative splicing. *J. Biol. Chem.* **270**: 958–965.
28. Kjellén, L., and Lindahl, U. (1991) Proteoglycans: Structures and interactions. *Annu. Rev. Biochem.* **60**: 443–475.
29. Knox, P. and Wells, P. (1979) Cell adhesion and proteoglycans. I. The effect of exogenous proteoglycans on the attachment of chick embryo fibroblasts to tissue culture plastic and collagen. *J. Cell Sci.* **40**: 77–88.
30. Kresse, H., Hausser, H., and Schönherr, E. (1993) Small proteoglycans. *Experientia* **49**: 403–416.
31. Levine, J.M. (1994) Increased expression of the NG2 chondroitin-sulfate proteoglycan after brain injury. *J. Neurosci.* **14**: 4716–4730.
32. Levine, J.M. and Card, J.P. (1987) Light and electron microscopic localization of a cell surface antigen (NG2) in rat cerebellum: Association with smooth protoplasmic astrocytes. *J. Neurosci.* **7**: 2711–2720.

33. Levine, J.M. and Stallcup, W.B. (1987) Plasticity of developing cerebellar cells *in vitro* studied with antibodies against the NG2 antigen. *J. Neurosci.* **7**: 2721–2731.
34. Lewandowska, K., Choi, H.U., Rosenberg, L.C., Zardi, L., and Culp, L.A. (1987) Fibronectin-mediated adhesion of fibroblasts: Inhibition by dermatan sulfate proteoglycan and evidence for a cryptic glycosaminoglycan-binding domain. *J. Cell Biol.* **105**: 1443–1454.
35. Margolis R.U. and Margolis R.K. (1994) The aggrecan-versican-neurocan family of proteoglycans. *Meth. Enzymol.* **245**: 105–126.
36. Margolis, R.K., Goossen, B., Tekotte, H., Hilgenberg, L., and Margolis, R.U. (1991) Effects of ß-xylosides on proteoglycan biosynthesis and morphology of PC12 pheochromocytoma cells and primary cultures of rat cerebellum. *J. Cell Sci.* **99**: 237–246.
37. Maurel, P., Rauch, U., Flad, M., Margolis, R.K., and Margolis, R.U. (1994) Phosphacan, a chondroitin sulfate proteoglycan of brain that interacts with neurons and neural cell adhesion molecules, is an extracellular variant of a receptor-type protein tyrosine phosphatase. *Proc. Natl. Acad. Sci. U. S. A.* **91**: 2512–2516.
38. Milev, P., Friedlander, D.R., Sakurai, T., Karthikeyan, L., Flad, M., Margolis, R.K., Grumet, M., and Margolis, R.U. (1994) Interactions of the chondroitin sulfate proteoglycan phosphacan, the extracellular domain of a receptor-type protein tyrosine phosphatase, with neurons, glia, and neural cell adhesion molecules. *J. Cell Biol.* **127**: 1703–1715.
39. Neame, P.J., Choi, H.U., and Rosenberg, L.C. (1989) The primary structure of the core protein of the small, leucine-rich proteoglycan (PG I) from bovine articular cartilage. *J. Biol. Chem.* **264**: 8653–8661.
40. Nishiyama, A., Dahlin, K.J., Prince, J.T., Johnstone, S.R., and Stallcup, W.B. (1991a) The primary structure of NG2, a novel membrane-spanning proteoglycan. *J. Cell Biol.* **114**: 359–371.
41. Nishiyama, A., Dahlin, K.J., and Stallcup, W.B. (1991b) The expression of NG2 proteoglycan in the developing rat limb. *Development* **111**: 933–944.
42. Nishiyama, A. and Stallcup, W.B. (1993) Expression of NG2 proteoglycan causes retention of type VI collagen on the cell surface. *Mol. Biol. Cell* **4**: 1097–1108.
43. Perris, R. and Johansson, S. (1987) Amphibian neural crest cell migration on purified extracellular matrix components: A chondroitin sulfate proteoglycan inhibits locomotion on fibronectin substrates. *J. Cell Biol.* **105**: 2511–2521.
44. Perris, R. and Johansson, S. (1990) Inhibition of neural crest cell migration by aggregating chondroitin sulfate proteoglycans is mediated by their hyaluronan-binding region. *Dev. Biol.* **137**: 1–12.
45. Rauch, U., Gao, P., Janetzko, A., Flaccus, A., Hilgenberg, L., Tekotte, H., Margolis, R.K., and Margolis, R.U. (1991) Isolation and characterization of developmentally-regulated chondroitin sulfate and chondroitin/keratan sulfate proteoglycans of brain identified with monoclonal antibodies. *J. Biol. Chem.* **266**: 14785–14801.
46. Rauch, U., Karthikeyan, L., Maurel, P., Margolis, R.U., and Margolis, R.K. (1992) Cloning and primary structure of neurocan, a developmentally regulated, aggregating chondroitin sulfate proteoglycan of brain. *J. Biol. Chem.* **267**: 19536–19547.

47. Reyes, A.A., Akeson, R., Brezina, L., and Cole, G.J. (1990) Structural requirements for neural cell adhesion molecule-heparin interaction. *Cell. Regul.* **1**: 567–576.
48. Rich, A.M., Pearlstein, E., Weissmann, G., and Hoffstein, S.T. (1981) Cartilage proteoglycans inhibit fibronectin-mediated adhesion. *Nature* **293**: 224–226.
49. Shinomura, T., Nishida, Y., Ito, K., and Kimata, K. (1993) cDNA Cloning of PG-M, a large chondroitin sulfate proteoglycan expressed during chondrogenesis in chick limb buds. Alternative spliced multiforms of PG-M and their relationships to versican. *J. Biol. Chem.* **268**: 14461–14469.
50. hitara, K., Yamada, H., Watanabe, K., Shimonaka, M., and Yamaguchi, Y. (1994) Brain-specific receptor-type protein-tyrosine phosphatase RPTPβ is a chondroitin sulfate proteoglycan *in vivo*. *J. Biol. Chem.* **269**: 20189–20193.
51. Stallcup, W.B. and Beasley, L. (1987) Bipotential glial precursor cells of the optic nerve express the NG2 proteoglycan. *J. Neurosci.* **7**: 2737–2744.
52. Stallcup, W.B., Dahlin, K., and Healy, P. (1990) Interaction of the NG2 chondroitin sulfate proteoglycan with type VI collagen. *J. Cell Biol.* **111**: 3177–3188.
53. Stanley, M.J., Liebersbach, B.F., Liu, W., Anhalt, D.J., and Sanderson, R.D. (1995) Heparan sulfate-mediated cell aggregation. Syndecans-1 and -4 mediate intercellular adhesion following their transfection into human B lymphoid cells. *J. Biol. Chem.* **270**: 5077–5083.
54. Torok, M.A., Evans, S.A.S., and Marcum, J.A. (1993) cDNA sequence for bovine biglycan (PGI) protein core. *Biochim. Biophys. Acta* **1173**: 81–84.
55. Verna, J-M., Fichard, A. and Saxod, R. (1989). Influence of glycosaminoglycans on neurite morphology and outgrowth patterns *in vitro*. *Int. J. Devl. Neurosci.* **7**: 389–399.
56. Winnemöller, M., Schmidt, G., and Kresse, H. (1991) Influence of decorin on fibroblast adhesion to fibronectin. *Eur. J. Cell Biol.* **54**: 10–17.
57. Winnemöller, M., Schön, P., Vischer, P., and Kresse, H. (1992) Interactions between thrombospondin and the small proteoglycan decorin: Interference with cell attachment. *Eur. J. Cell Biol.* **59**: 47–55.
58. Yamagata, M., Suzuki, S., Akiyama, S.K., Yamada, K.M., and Kimata, K. (1989) Regulation of cell-subrate adhesion by proteoglycans immobilized on extracellular substrates. *J. Biol. Chem.* **264**: 8012–8018.
59. Yamaguchi, Y. and Ruoslahti, E. (1988) Expression of human proteoglycan in Chinese hamster ovary cells inhibits cell proliferation. *Nature* **336**: 244–246.
60. Yamaguchi, Y., Mann, D.M. and Ruoslahti, E. (1990) Negative regulation of transforming growth factor-b by the proteoglycan decorin. *Nature* **346**: 281–284.
61. Zako, M., Shinomura, T., Ujita, M., Ito, K., and Kimata, K. (1995) Expression of PG-M(V3), an alternatively spliced form of PG-M without a chondroitin sulfate attachment region in mouse and human tissues. *J. Biol. Chem.* **270**: 3914–3918.
62. Zimmermann, D.R. and Ruoslahti, E. (1989) Multiple domains of the large fibroblast proteoglycan, versican. *EMBO J.* **8**: 2975–2981.

INDEX